Texts in Applied Mathematics **8**

Texts in Applied Mathematics

James B. Seaborn

Hypergeometric Functions and Their Applications

With 59 Illustrations

Springer-Verlag
New York Berlin Heidelberg London
Paris Tokyo Hong Kong Barcelona

James B. Seaborn
Department of Physics
University of Richmond
VA 23173
USA

Editors

F. John
Courant Institute of
  Mathematical Sciences
New York University
New York, NY 10012
USA

J.E. Marsden
Department of
  Mathematics
University of California
Berkeley, CA 94720
USA

L. Sirovich
Division of Applied
  Mathematics
Brown University
Providence, RI 02912
USA

M. Golubitsky
Department of
  Mathematics
University of Houston
Houston, TX 77004
USA

W. Jäger
Department of Applied
  Mathematics
Universität Heidelberg
Im Neuenheimer Feld 294
6900 Heidelberg, FRG

Mathematics Subject Classification: 33-01, 33A30, 42CXX

Library of Congress Cataloging-in-Publication Data
Seaborn, James B.
    Hypergeometric functions and their applications / by James B.
Seaborn.
        p.   cm.
    Includes bibliographical references and index.
    ISBN 0-387-97558-6
    1. Functions, Hypergeometric.   2. Mathematical physics.
I. Title.
QA353.H9S43   1991
515'.55—dc20   91-18910

Printed on acid-free paper.

Photocomposed from a LaTex file.
Printed and bound by R.R. Donnelley & Sons, Harrisonburg, VA.
Printed in the United States of America.

9 8 7 6 5 4 3 2 1

ISBN 0-387-97558-6 Springer-Verlag New York Berlin Heidelberg
ISBN 3-540-97558-6 Springer-Verlag Berlin Heidelberg New York

To Gwen

Series Preface

Mathematics is playing an ever more important role in the physical and biological sciences, provoking a blurring of boundaries between scientific disciplines and a resurgence of interest in the modern as well as the classical techniques of applied mathematics. This renewal of interest, both in research and teaching, has led to the establishment of the series: *Texts in Applied Mathematics (TAM)* .

The development of new courses is a natural consequence of a high level of excitement on the research frontier as newer techniques, such as numerical and symbolic computer systems, dynamical systems, and chaos, mix with and reinforce the traditional methods of applied mathematics. Thus, the purpose of this textbook series is to meet the current and future needs of these advances and encourage the teaching of new courses.

TAM will publish textbooks suitable for use in advanced undergraduate and beginning graduate courses, and will complement the *Applied Mathematical Sciences (AMS)* series, which will focus on advanced textbooks and research level monographs.

Preface

A wide range of problems exists in classical and quantum physics, engineering, and applied mathematics in which special functions arise. The procedure followed in most texts on these topics (e.g., quantum mechanics, electrodynamics, modern physics, classical mechanics, etc.) is to formulate the problem as a differential equation that is related to one of several special differential equations (Hermite's, Bessel's, Laguerre's, Legendre's, etc.). The corresponding special functions are then introduced as solutions with some discussion of recursion formulas, orthogonality relations, asymptotic expressions, and other properties as appropriate. In every instance, the reader is referred to a standard text on applied mathematical methods for more detail. This is all very reasonable and proper.

Each special function can be defined in a variety of ways and different authors choose different definitions (Rodrigues formulas, generating functions, contour integrals, etc.). Whatever the starting definition, it is usually shown to be expressible as a series, because this is frequently the most practical way to obtain numerical values for the functions. Also it is often shown—or at least stated—that the special function can be expressed in terms of some generalized hypergeometric function.

In this book, we follow a different track. Each special function arises in one or more physical contexts as a solution of a differential equation that can be transformed into the hypergeometric equation (or its confluent form). The special function is then *defined* in terms of a generalized hypergeometric function. From this definition, many of the interesting and important properties encountered in standard upper level textbooks (recursion formulas, the generating function, orthogonality relations, Rodrigues formula, asymptotic expressions, and various series and integral representations) are derived and the equivalence of this definition to other definitions is established.

This approach is interesting, and it is instructive to see that most of the special functions encountered in applied mathematics have a common root in their relation to the hypergeometric function. The reader may notice that derivations are not always carried out in the simplest or the most straightforward manner. This is usually intentional—a consequence of a deliberate choice made in favor of furnishing the clearest and most direct connections between the functions of applied mathematics and the hypergeometric function rather than of finding the most elegant path to a given result.

In most cases, I have not introduced a mathematical topic until it is necessary for the discussion. For example, complex analysis is not really needed until we begin to look at alternate forms and integral representations of the special functions in Chapter 9. So I have deferred this subject until Chapters 7 and 8 with a few appropriate reminders along the way that it is coming.

Also concerning the mathematics, I have reviewed some of the fundamental notions from calculus (function, continuity, convergence, etc.), which a student may not remember very clearly, but I have tried to hold this kind of review to a minimum. In a few instances, to provide some insight without getting too heavily bogged down in the mathematics, I have used heuristic arguments to establish a desired result.

The range of this book is intentionally rather narrow. There are many interesting and useful topics (conformal mapping, Sturm–Liouville theory, Green's functions, to name a few) which are related to those I have discussed, but which are outside the scope of the task I have in mind.

In writing this book I have assumed that the reader has completed two or three semesters of calculus and has some knowledge of Schrödinger's equation (perhaps, from a course in modern physics or an introductory course in quantum mechanics). Courses at the intermediate level in classical mechanics and electromagnetism are also desirable, but not essential. The book should be accessible to a reader with this minimum preparation. A student who has completed the intermediate courses in an undergraduate physics or engineering curriculum would have a much greater appreciation for the subject matter treated here.

This is all well-plowed ground and I am grateful to those who have worked these fields before me. There are many excellent books on mathematical analysis and methods of mathematical physics, and I have profited greatly from a number of them. Listed in the Bibliography are those that have been most helpful to me in writing this book.

I should like to express my deep gratitude to Professor Gerald Speisman who first excited my interest in this subject a long time ago. Finally, I wish to thank the anonymous reviewers for very helpful comments and suggestions.

Contents

Symbol Index

1

Special Functions in Applied Mathematics

Certain mathematical functions occur often enough in fields like physics and engineering to warrant special consideration. They form a class of well-studied functions with an extensive literature and, appropriately enough, are collectively called *special functions*. These functions carry such names as Bessel functions, Laguerre functions, and the like. Most of the special functions encountered in such applications have a common root in their relation to the *hypergeometric function*. The purpose of this book is to establish this relationship and use it to obtain many of the interesting and important properties of the special functions met in applied mathematics.

These matters are explored soon enough, but first let us recall some basic mathematical concepts from calculus.

1.1   Variables, Functions, Limits, and Continuity[1]

Let a and b represent two real numbers with $a < b$. Now let x be any real number in the interval between a and b. We say that x is a *variable* defined in the interval $a \leq x \leq b$. If one or more values of a second variable y are determined for each value of x in this interval, then y *is a function of* x in the interval and we denote this functional relationship by

$$y = f(x).$$

If there is only one value of y for each value of x, then y is a *single-valued* function of x.

The *limit of a function* $f(x)$ implies that as x gets closer and closer to some value x_0, $f(x)$ gets closer and closer to a value y_0. In more precise mathematical language, given any number $\varepsilon > 0$, there exists a number $\delta > 0$ such that[2]

$$|f(x) - y_0| < \varepsilon \quad \text{when} \quad |x - x_0| < \delta.$$

---

[1]For more detail see, for example, W. Kaplan, *Advanced Calculus*, Addison-Wesley, Reading, MA, 1953.

[2]We use the notation $|x|$ to denote the *absolute value of* x.

Formally we express this notion of a limit of a function by

$$\lim_{x \to x_0} f(x) = y_0.$$

If this limit is independent of the direction from which x approaches x_0 (i.e., through values greater than x_0 or through values less than x_0), then we say that $f(x)$ is *continuous* at x_0. Furthermore, $f(x)$ is continuous over the interval $a < x < b$ if it is continuous at each point x_0 in the interval $a < x_0 < b$.

These fundamental notions[3] of a function of a variable and its continuity and limit will appear often in our study of the special functions and their relation to the hypergeometric functions.

1.2   Why Study Special Functions?

The sine function is so familiar that we do not think of it as being particularly *special*. Nevertheless, it serves to illustrate a point. Consider two different physical phenomena: (a) the behavior of an elastically bound particle and (b) the spatial variation of the magnetic field in a resonant cavity.

a. *Elastically bound particle in one dimension.* The position x of a particle of mass m that moves under the influence of a force

$$F = -m\omega^2 x \tag{1.1}$$

is given at time t by[4]

$$x = b\sin(\omega t + \phi) \tag{1.2}$$

where b and ϕ are constants.

b. *Magnetic field in a resonant cavity.*[5] Maxwell's theory of electrodynamics predicts that the magnetic field B in a cyclindrical resonant cavity of length d varies with distance x along the axis of the cavity according to

$$B = B_0 \sin\left[\frac{n\pi x}{d}\right] \qquad n = 1, 2, 3, \ldots.$$

Here are two physically unrelated situations. In both, the behavior of the physical system is described by the *same* mathematical function, the sine. Such a recurrence of the same mathematical function in widely different contexts is encountered over and over again in the study of mathematical physics, as we shall soon see.

---

[3] Appropriate generalizations to functions of a complex variable are considered in Chapter 7.

[4] See p. 3 *ff.*

[5] J.D. Jackson, *Classical Electrodynamics*, Wiley, New York, 1962, p. 253.

1.3 Special Functions and Power Series

A special function can be represented as a power series.[6] In fact, it may be defined in this way. For example,

$$\sin\theta = \frac{\theta}{1} - \frac{\theta^3}{3!} + \frac{\theta^5}{5!} - \cdots = \sum_{k=0}^{\infty} \frac{(-1)^k \theta^{2k+1}}{(2k+1)!}, \tag{1.3}$$

where the factorial function[7] is defined by $n! = 1 \cdot 2 \cdot 3 \cdot 4 \cdots n$. Similarly, for the cosine function

$$\cos\theta = 1 - \frac{\theta^2}{2!} + \frac{\theta^4}{4!} - \cdots = \sum_{k=0}^{\infty} \frac{(-1)^k \theta^{2k}}{(2k)!}. \tag{1.4}$$

Why should the mathematical functions which arise in such applications be expressible as power series? Let us see. The behavior of a physical system is commonly represented by a differential equation. For example, the motion of a particle of mass m moving in one dimension is described by Newton's second law of motion,

$$F = m\frac{d^2 x}{dt^2}, \tag{1.5}$$

which is a second-order differential equation in x (position) and t (time). In general, the force F depends on x and t. Suppose the particle obeys Hooke's law. In this case, the force is given by Eq. (1.1). The parameter ω is a constant. We then have from Eq. (1.5),

$$\frac{d^2 x}{dt^2} + \omega^2 x = 0. \tag{1.6}$$

The problem is to solve this equation for x as a function of t. One very powerful method for solving differential equations is to assume a power series solution. By this we mean a solution of the form

$$x(t) = \sum_{n=0}^{\infty} a_n t^{s+n},$$

where the a_n are constant coefficients to be determined.[8] Substitution of this series for x in Eq. (1.6) yields

$$\sum_{n=0}^{\infty} a_n(s+n)(s+n-1)t^{s+n-2} + \omega^2 \sum_{n=0}^{\infty} a_n t^{s+n} = 0.$$

---

[6] A discussion of the *convergence* and other properties of infinite series is given in Chapter 2.

[7] The function $n!$ is called n *factorial*.

[8] Some authors refer to this procedure as the *method of Frobenius*. The term appears to have a more restricted meaning as used by H. Jeffreys and B.S. Jeffreys, *Methods of Mathematical Physics*, Cambridge University Press, Cambridge, 1956, p. 482.

In the first sum in this equation we redefine the summation index by $n' = n - 2$ which gives

$$a_0 s(s-1)t^{s-2} + a_1(s+1)st^{s-1} + \sum_{n'=0}^{\infty} a_{n'+2}(s+n'+2)(s+n'+1)t^{s+n'}$$

$$+ \omega^2 \sum_{n=0}^{\infty} a_n t^{s+n} = 0.$$

Since n' is a dummy index, we can drop the prime and combine the two sums to get

$$a_0 s(s-1)t^{s-2} + a_1(s+1)st^{s-1}$$

$$+ \sum_{n=0}^{\infty} \left[(s+n+2)(s+n+1)a_{n+2} + \omega^2 a_n\right]t^{s+n} = 0.$$

This equation holds for all values of t, hence the coefficients of the different powers of t must all vanish separately. That is,

$$s(s-1)a_0 = 0, \tag{1.7a}$$

$$(s+1)sa_1 = 0, \tag{1.7b}$$

$$(s+n+2)(s+n+1)a_{n+2} + \omega^2 a_n = 0. \tag{1.7c}$$

The first two of these equations are satisfied if we take $s = 0$ in which case a_0 and a_1 are arbitrary. With this choice of s the third equation yields the two sets of coefficients

$$a_n = \frac{(-1)^{n/2}\omega^n}{n!} a_0, \qquad n = \text{even},$$

and

$$a_n = \frac{(-1)^{(n-1)/2}\omega^{n-1}}{n!} a_1, \qquad n = \text{odd}.$$

The general solution of Eq. (1.6) is then

$$x(t) = a_0 \sum_{\substack{n=0 \\ n=even}}^{\infty} \frac{(-1)^{n/2}\omega^n t^n}{n!} + a_1 \sum_{\substack{n=1 \\ n=odd}}^{\infty} \frac{(-1)^{(n-1)/2}\omega^{n-1}t^n}{n!}.$$

On redefining the summation indices we have

$$x(t) = a_0 \sum_{k=0}^{\infty} \frac{(-1)^k(\omega t)^{2k}}{(2k)!} + \frac{a_1}{\omega} \sum_{k=0}^{\infty} \frac{(-1)^k(\omega t)^{2k+1}}{(2k+1)!}.$$

The index in each of the sums now runs over *all* nonnegative integers.

For a given value of ωt each of the two series in this equation represents a number that, in practice, can be obtained by evaluating the series numerically. We also notice that if we let $\omega t = \theta$, then these are the same series we have already identified in Eqs. (1.3) and (1.4). Thus, the solution of Eq. (1.6) may be written

$$x(t) = a_0 \cos \omega t + (a_1/\omega) \sin \omega t.$$

The general solution of the second-order differential equation is the sum of two linearly independent solutions (a sine function and a cosine function).[9] We can rewrite this last form of the solution as

$$x(t) = A \cos \omega t + B \sin \omega t \qquad (1.8)$$

where we have redefined the arbitrary constants of integration. These two constants, A and B, are determined from the initial conditions. For example, suppose that at time $t = 0$ the particle is at the point x_0 with a velocity v_0. With $t = 0$ Eq. (1.8) becomes

$$x(0) = A = x_0.$$

On differentiating Eq. (1.8) and evaluating at $t = 0$, we get for the velocity

$$\left.\frac{dx}{dt}\right|_{t=0} = \omega B = v_0.$$

Using these values of A and B, rewrite Eq. (1.8) as

$$x(t) = x_0 \cos \omega t + (v_0/\omega) \sin \omega t. \qquad (1.9)$$

With a further redefinition of the constants of integration this solution can be written in the form of Eq. (1.2) where

$$b = \sqrt{x_0^2 + (v_0/\omega)^2} \quad \text{and} \quad \phi = \tan^{-1}(x_0\omega/v_0). \qquad (1.10)$$

We have illustrated here the treatment of a typical problem in mathematical physics: Set up the differential equation describing the behavior of the system, solve the differential equation, and apply the physical boundary conditions to find the constants of integration.

The remarkable thing is that surprisingly often these differential equations can be related to the same fundamental equation, and a wide range of phenomena can be described by a mere handful of special mathematical functions. In this book it is our aim to demonstrate this unity and to examine the properties of the special functions that arise.

---

[9]For a definition of *linear independence*, see p. 28.

1.4   The Gamma Function: Another Example from Physics

Special functions may also occur in applied mathematics in ways other than as solutions of differential equations. To illustrate this point let us consider the distribution of speeds of free particles in a gas at temperature T.

From statistical mechanics Maxwell deduced that for a gas in equilibrium the fraction of particles (of mass m) with speeds between v and $v + dv$ is[10]

$$g(v)dv = Cv^2 e^{-mv^2/2kT} dv$$

where k is Boltzmann's constant. Since every particle must have *some* speed, the fraction with speeds between zero and infinity is one. That is,

$$\int_0^\infty g(v)\, dv = 1. \tag{1.11}$$

In order to find the constant C in the distribution above, we must evaluate the integral

$$\int_0^\infty v^2 e^{-mv^2/2kT} dv.$$

Let us start by changing the variable of integration to $t = mv^2/2kT$. Equation (1.11) then becomes

$$\frac{1}{2}C\left(\frac{2kT}{m}\right)^{3/2}\int_0^\infty t^{3/2-1} e^{-t} dt = 1. \tag{1.12}$$

The integral here is a particular example of an integral representation of a special function, which we define for $z > 0$ by

$$\Gamma(z) = \int_0^\infty t^{z-1} e^{-t} dt. \tag{1.13}$$

This very important function is called the *gamma function*.[11]

1.4.1   PROPERTIES OF THE GAMMA FUNCTION

The gamma function crops up repeatedly in applied mathematical analysis. It is convenient here to examine some of its properties which are extremely useful later in this book. An integration by parts in Eq. (1.13) yields the property

$$\Gamma(z+1) = z\Gamma(z). \tag{1.14}$$

---

[10]For example, see F. Reif, *Fundamentals of Statistical and Thermal Physics*, McGraw-Hill, New York, 1965, p. 267.

[11]N.N. Lebedev, *Special Functions and Their Applications*, Dover, New York, 1972, p. 1.

From this, it is easily seen that if z is a nonnegative integer, then the gamma function is identical with the factorial function,

$$\Gamma(n+1) = n! \tag{1.15}$$

with $\Gamma(1) = 0! = 1$.

On replacing t by ax $(a > 0)$ and z by $z + 1$ in Eq. (1.13) we obtain the formula

$$\int_0^\infty x^z e^{-ax} dx = \frac{\Gamma(z+1)}{a^{z+1}}, \tag{1.16}$$

where a is independent of x.

By the change of variable $s = e^{-t}$ in Eq. (1.13) we get another integral representation of the gamma function,[12]

$$\Gamma(z) = \int_0^1 (\log s^{-1}))^{z-1} ds. \tag{1.17}$$

A very useful formula for a product of two gamma functions is readily obtained from the defining integral, Eq. (1.13). We have,

$$\Gamma(x+y) = \int_0^\infty e^{-t} t^{x+y-1} dt,$$

where x and y are both positive. By changing the variable of integration to $s = t/(1 + p)$ with $p > -1$, we can rewrite this integral as

$$\Gamma(x+y) = (1+p)^{x+y} \int_0^\infty e^{-(1+p)s} s^{x+y-1} ds.$$

Now we multiply both sides of this equation by $p^{x-1}/(1 + p)^{x+y}$ and integrate[13] with respect to p to obtain,

$$\Gamma(x+y) \int_0^\infty p^{x-1} (1+p)^{-x-y} dp = \int_0^\infty e^{-s} s^{x+y-1} \left(\int_0^\infty e^{-ps} p^{x-1} dp \right) ds. \tag{1.18}$$

We see from Eq. (1.16) that the integral in parentheses is just $s^{-x}\Gamma(x)$, so that the right hand side of Eq. (1.18) now becomes

$$\Gamma(x) \int_0^\infty e^{-s} s^{y-1} ds = \Gamma(x)\Gamma(y).$$

Thus, Eq. (1.18) may be written as

$$\Gamma(x)\Gamma(y) = \Gamma(x+y) \int_0^\infty p^{x-1} (1+p)^{-x-y} dp. \tag{1.19}$$

---

[12] We denote the *natural* logarithm of a number x by $\log x$.

[13] In this book, we assume that it is permissible to interchange orders of integration in repeated integrals. For a further discussion, see E.C. Titchmarsh, *The Theory of Functions*, Oxford University Press, Oxford, 1939, p. 53.

As a special case we set $y = 1 - x$ to obtain the formula

$$\Gamma(x)\Gamma(1 - x) = \int_0^\infty \frac{p^{x-1}}{1 + p} \, dp, \tag{1.20}$$

which is valid for $0 < x < 1$.

The integral in Eq. (1.19) may be further transformed by a change of variable $t = p/(1 + p)$,

$$\int_0^\infty p^{x-1}(1 + p)^{-x-y} dp = \int_0^1 t^{x-1}(1 - t)^{y-1} dt. \tag{1.21}$$

With this result inserted into Eq. (1.19) the product of the gamma functions may be expressed as

$$\Gamma(x)\Gamma(y) = \Gamma(x + y) \int_0^1 t^{x-1}(1 - t)^{y-1} dt. \tag{1.22}$$

Consistent with the limits of integration we set $t = \cos^2 \theta$ to obtain

$$\Gamma(x)\Gamma(y) = 2\Gamma(x + y) \int_0^{\pi/2} (\cos \theta)^{2x-1}(\sin \theta)^{2y-1} d\theta. \tag{1.23}$$

On redefining the parameters according to $\mu = x - \frac{1}{2}$ and $\nu = y - \frac{1}{2}$, we may rewrite Eq. (1.23) as

$$\Gamma\left(\mu + \tfrac{1}{2}\right)\Gamma\left(\nu + \tfrac{1}{2}\right) = 2\Gamma(\mu + \nu + 1) \int_0^{\pi/2} (\cos \theta)^{2\mu}(\sin \theta)^{2\nu} d\theta. \tag{1.24}$$

If we set $\mu = \nu = 0$, this equation reduces to

$$\Gamma\left(\tfrac{1}{2}\right)\Gamma\left(\tfrac{1}{2}\right) = 2\Gamma(1) \int_0^{\pi/2} d\theta.$$

Thus,

$$\Gamma\left(\tfrac{1}{2}\right) = \sqrt{\pi}. \tag{1.25}$$

Another important result[14] also follows from the formula for the product of two gamma functions. With $x = y = z$ Eq. (1.22) becomes

$$\Gamma(z)\Gamma(z) = \Gamma(2z) \int_0^1 t^{z-1}(1 - t)^{z-1} dt.$$

By changing the variable of integration to $u = 2t - 1$ we get

$$\Gamma(z)\Gamma(z) = 2^{1-2z}\Gamma(2z) \int_{-1}^1 (1 - u^2)^{z-1} du.$$

---

[14]G. Arfken, *Mathematical Methods for Physicists*, Academic Press, New York, 1970, p. 468.

Since the integrand is an even function of u, we can write

$$\Gamma(z)\Gamma(z) = 2^{2-2z}\Gamma(2z)\int_0^1 (1-u^2)^{z-1}du.$$

With another variable change $v = u^2$ the integral takes the form

$$\Gamma(z)\Gamma(z) = 2^{1-2z}\Gamma(2z)\int_0^1 v^{-1/2}(1-v)^{z-1}dv.$$

Now we evaluate the integral by means of Eq. (1.22) to obtain the formula

$$\Gamma(z)\Gamma\left(z+\tfrac{1}{2}\right) = \sqrt{\pi}\,2^{1-2z}\Gamma(2z). \tag{1.26}$$

This relation is called the *duplication formula* for the gamma function.

All of the formulas in this section will prove quite useful to us later in this book.

1.4.2   VELOCITY DISTRIBUTION IN AN IDEAL GAS

Returning now to Eq. (1.12), we see that the integral is

$$\int_0^\infty t^{3/2-1}e^{-t}dt = \tfrac{1}{2}\sqrt{\pi}, \tag{1.27}$$

where we have used Eqs. (1.14) and (1.25). With this result, we can evaluate the normalization constant in Eq. (1.12),

$$C = \frac{4}{\sqrt{\pi}}\left(\frac{m}{2kT}\right)^{3/2}.$$

Finally, the distribution of particle speeds in an ideal gas is given by,

$$g(v) = 4\pi\left(\frac{m}{2\pi kT}\right)^{3/2}e^{-mv^2/2kT}v^2. \tag{1.28}$$

The fraction of particles in this gas with speeds less than or equal to some value v_1 is

$$f_{v\le v_1} = \int_0^{v_1} g(v)\,dv.$$

This involves evaluation of an integral of the form

$$\gamma(z,b) = \int_0^b e^{-t}t^{z-1}dt,$$

which is known as the *incomplete gamma function*.

1.5   A Look Ahead

Many techniques exist for evaluating definite integrals. One very powerful
tool is contour integration, which is discussed in Chapter 8. The value
for $\Gamma\left(\frac{1}{2}\right)$ in Eq. (1.25) may be obtained by this method.[15] A geometrical
approach to evaluating $\Gamma\left(\frac{1}{2}\right)$ is left as an exercise.[16]

Mathematical functions like the sine, the cosine, and the exponential,
along with polynomials and other algebraic expressions, are referred to
as *elementary functions*. They will occur frequently in our work here as
we have already seen, but the focus of our study will be on the group of
special functions called *higher transcendental functions* of which the gamma
function is one example.

EXERCISES

1. Start with Eq. (1.2) and obtain an expression for x in the form of Eq.
   (1.9). From your result write b and ϕ explicitly in terms of x_0 and v_0.

2. Follow the series method illustrated in the text to show that the
   solution to the differential equation,

$$\frac{du(z)}{dz} - u(z) = 0,$$

   is proportional to the exponential function,

$$e^z = \sum_{k=0}^{\infty} \frac{z^k}{k!}.$$

   Note that since the differential equation is first order, there is only
   one arbitrary constant of integration.

3. Two more solutions to Eq. (1.6) may be obtained by taking different
   values for s in Eqs. (1.7).

   a. If $s = 1$, then Eq. (1.7b) requires that $a_1 = 0$.
   b. If $s = -1$, then Eq. (1.7a) requires that $a_0 = 0$.

   Show that these values of s lead to the solutions $x = A\sin\omega t$ and
   $x = B\cos\omega t$, respectively. Express the constants A and B explicitly
   in terms of a_0 and a_1.

---

[15]See Eq. (9.3) with $z = \frac{1}{2}$.
[16]See Exercise 1.12.

4. A particle of mass m is traveling in a straight line at a constant speed v_0. See Figure 1.1. At time t_0 it suddenly enters a viscous medium in which it experiences a retarding force proportional to its speed, i.e.,

$$F = -m\gamma v,$$

where γ is a constant. With velocity as the dependent variable write down the equation of motion (Newton's second law) for the particle in the medium. Use the series method to solve this first-order differential equation for velocity as a function of time. Express your result in terms of elementary functions and the given parameters. By direct integration of your result find the distance the particle has penetrated into the medium at time $t_1 > t_0$.

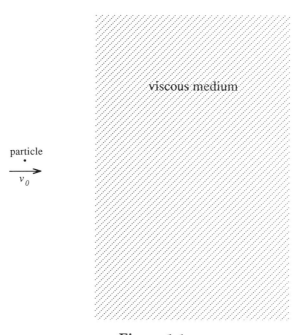

Figure 1.1

5. A solution to the second-order linear differential equation

$$zu''(z) + u'(z) + zu(z) = 0$$

has the form

$$u(z) = \sum_{k=0}^{\infty} c_k z^{k+s},$$

where s is a number. Obtain the general recursion formula for the coefficients c_k. Can both c_0 and c_1 be zero at the same time? Explain.

What values may the number s have? Under what conditions? For each value of s obtain the explicit expression for the coefficient c_k. Express your result in terms of factorials. Write down the explicit series solution corresponding to each value of s. Show that the two solutions are not independent by showing that one reduces to the other (except possibly for an overall multiplicative constant).

6. Obtain the recurrence formula

$$\Gamma(z+1) = z\Gamma(z)$$

from Eq. (1.13) as indicated in the text. Show that if z is a nonnegative integer $(z = n)$, then

$$\Gamma(n+1) = n!\,.$$

Integrate Eq. (1.13) directly to show that $\Gamma(1) = 1$.

7. By repeated application of the property

$$\Gamma(z+1) = z\Gamma(z),$$

show that for any nonnegative integer n

$$\Gamma(2n+2) = \frac{2^{2n+1}n!}{\sqrt{\pi}} \Gamma\left(n + \tfrac{3}{2}\right).$$

Show that this result also follows from the duplication formula.

8. Carry out the indicated changes of variable to get Eqs. (1.16) and (1.17).

9. By making an appropriate change of variable of integration show that

$$\int_0^\infty x^q e^{-bx^p} dx = \frac{\Gamma\left(\frac{q+1}{p}\right)}{pb^{(q+1)/p}},$$

where p, q, and b are constants.

10. Show that for any positive integer n

$$\Gamma(z) = \frac{\Gamma(z+n+1)}{z(z+1)(z+2)\cdots(z+n)}.$$

11. With an appropriate change of variable in the integral in Eq. (1.19) obtain the result

$$\int_0^\infty p^{x-1}(1+p)^{-x-y} dp = \int_0^1 t^{x-1}(1-t)^{y-1} dt.$$

See Eq. (1.21).

12. In Eq. (1.13) make the change of variable $t = r^2$ and take $z = \frac{1}{2}$ so that

$$\Gamma\left(\tfrac{1}{2}\right) = 2 \int_0^\infty e^{-r^2} dr.$$

Let r and θ be the polar coordinates of a point in the x-y plane with $x = r\cos\theta$ and $y = r\sin\theta$ (see Figure 1.2), and integrate the function

$$f(x, y) = e^{-(x^2+y^2)}$$

over the entire first quadrant to obtain the result in Eq. (1.25).

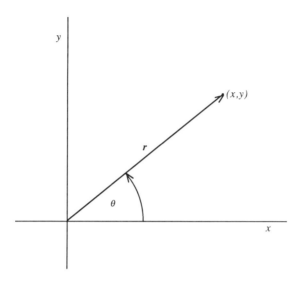

Figure 1.2

13. With changes of variable similar to that in the previous problem show that

$$\Gamma(x)\Gamma(y) = 4 \int_0^\infty \int_0^\infty e^{-(p^2+q^2)} p^{2x-1} q^{2y-1} \, dp \, dq.$$

Now transform to polar coordinates according to $p = r\cos\theta$ and $q = r\sin\theta$ and carry out this integration over the entire first quadrant to obtain Eq. (1.23).

14. From the definition of the gamma function derive the formula

$$\int_0^\infty e^{-a^2 y^2} y^b \, dy = \frac{\Gamma\left(\frac{b+1}{2}\right)}{2a^{b+1}}.$$

Verify that this is consistent with the result in Exercise 1.9. Show that for any nonnegative integer n,

$$\int_0^\infty x^{2n} e^{-cx^2} \, dx = \frac{1}{2^{2n+1}} \frac{(2n)!}{n!} \left(\frac{1}{c}\right)^{n+(1/2)} \sqrt{\pi},$$

and
$$\int_0^\infty x^{2n+1} e^{-cx^2} dx = \frac{n!}{2c^{n+1}}.$$

15. For any distribution $P(x)$ of some quantity x over a system, the average value of any other quantity f which is a function of x is

$$\overline{f(x)} = \int P(x)f(x)\,dx.$$

Use this property of distributions to find the average speed and the average nonrelativistic kinetic energy of a particle in a dilute gas where the distribution of speeds is described by Eq. (1.28). Consider only those particles whose kinetic energies at a given instant do not exceed some energy E and find an expression for the kinetic energy per particle averaged over this fraction of the distribution.

16. Show that for $p > -1$

$$\int_0^{\pi/2} \cos^p \theta \, d\theta = \int_0^{\pi/2} \sin^p \theta \, d\theta = \frac{\sqrt{\pi}\,\Gamma\left(\frac{p+1}{2}\right)}{2\Gamma\left(\frac{p}{2} + 1\right)}.$$

17. By making a suitable change of variable show that for any nonnegative integer n,

$$\int_{-1}^1 (1 - x^2)^n dx = \frac{2^{2n+1}(n!)^2}{(2n+1)!}.$$

18. Let ν and b be positive numbers. Derive the formula

$$\int_0^b \frac{dx}{\sqrt{b^\nu - x^\nu}} = b\sqrt{\frac{\pi}{b^\nu}} \frac{\Gamma\left(\frac{1}{\nu} + 1\right)}{\Gamma\left(\frac{1}{\nu} + \frac{1}{2}\right)}.$$

19. Show that
$$\int_0^\infty \frac{dt}{t^{1/2}(1 + t)} = \pi.$$

20. An integral which has applications in physics is defined by

$$\operatorname{erf} z = \int_0^z e^{-u^2} du.$$

It is called the *error function*.[17] because of its importance in the theory of errors. Show that the error function can be expressed as an incomplete gamma function.

---

[17] With a different normalization it is known as the *probability integral*. See, for example, N.N. Lebedev, *op. cit.*, p. 16. Other authors use different nomenclature and different normalization.

2

Differential Equations and Special Functions

2.1 Infinite Series

We are mainly interested in those special functions that appear as solutions of differential equations. As suggested in Section 1.3, the solutions can be expressed as infinite series. For this reason, it is convenient here to review some general properties of such series.[1]

In cases of physical interest we require that the series *converge*. To make clear what we mean by convergence of an infinite series we consider the sum of the first n terms of the series, which we denote by

$$S_n = \sum_{k=1}^{n} u_k,$$

and which we call the nth *partial sum*.

If the sequence of partial sums converges to a finite limit S according to

$$\lim_{n \to \infty} S_n = S,$$

then we say that the infinite series *converges* and has the value S. That is,

$$\sum_{k=1}^{\infty} u_k = S.$$

If the terms u_k vary in sign, then some partial cancellation occurs and convergence is much more likely. If the series

$$\sum_{k=1}^{\infty} |u_k|$$

converges, then we say that the series

$$\sum_{k=1}^{\infty} u_k$$

---

[1]For a more complete discussion see W. Kaplan, *op. cit.* or E.C. Titchmarsh, *op. cit.*

is *absolutely convergent*. The convergence of a particular series may be established according to one of several convergence tests.[2]

Of more interest to us is an infinite series of functions $u_k(x)$ of some variable x. Consider the series

$$S(x) = \sum_{k=1}^{\infty} u_k(x)$$

for x in the interval $a \leq x \leq b$. The series converges if, given any x in the interval and any positive number ε, an n exists such that the absolute difference between the series and the nth partial sum is less than ε,

$$|S(x) - S_n(x)| < \varepsilon \quad \text{with} \quad a \leq x \leq b.$$

We take n to be the smallest integer for which this holds.

Notice that in general n depends on ε and on x. It may happen that for each ε another number N exists that is independent of x such that $n(x, \varepsilon) < N(\varepsilon)$ for all x in the interval. In this case we say that the series *converges uniformly* in the interval.

Infinite series that converge uniformly have some very useful properties, which we now list without proof.[3]

  a. The sum of a uniformly convergent series of continuous functions is continuous.

  b. A uniformly convergent series of continuous functions can be integrated term by term.

  c. A uniformly convergent series of continuous functions can be differentiated term by term if the terms all have continuous derivatives and the series of derivatives converges uniformly.

For the series considered in this book we assume that all of these properties hold unless explicitly stated to the contrary.

2.2 Analytic Functions

Generally, the functions we encounter in physics and applied mathematics are *analytic*. To say that a function $f(x)$ is analytic for values of the variable x in the range $a < x < b$ we mean that for each point x_0 in this interval $f(x)$ can be written as a power series,

$$f(x) = \sum_{n=0}^{\infty} c_n(x - x_0)^n, \tag{2.1}$$

---

[2]See, for example, W. Kaplan, *op. cit.* or E.C. Titchmarsh, *op. cit.*
[3]See E.C. Titchmarsh, *op. cit.*

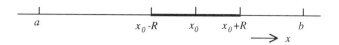

Figure 2.1

where the numbers c_n are independent of x. In other words, $f(x)$ is analytic in the interval if and only if the sum converges for every point x_0 in the interval $a < x_0 < b$ and x is in the range $x_0 - R < x < x_0 + R$, where the nonnegative number R is called the *radius of convergence* (see Figure 2.1). Both the coefficients c_n and the radius of convergence R depend on x_0, but *not* on x.

On differentiating the sum in Eq. (2.1) term by term we get

$$f'(x) = c_1 + 2c_2(x - x_0) + 3c_3(x - x_0)^2 + 4c_4(x - x_0)^3 + \cdots .$$

A second differentiation gives

$$f''(x) = 1 \cdot 2c_2 + 2 \cdot 3c_3(x - x_0) + 3 \cdot 4c_4(x - x_0)^2 + 4 \cdot 5(x - x_0)^3 + \cdots .$$

By continuing with successive differentiations we obtain for the nth derivative

$$f^{(n)}(x) = n!c_n + \frac{(n+1)!}{1!}c_{n+1}(x - x_0)^1 + \frac{(n+2)!}{2!}c_{n+2}(x - x_0)^2 + \cdots$$
$$+ \frac{(n+k)!}{k!}c_{n+k}(x - x_0)^k + \cdots .$$

If we set $x = x_0$ in this last equation, we find that for the nth derivative we have

$$f^{(n)}(x_0) = n!c_n,$$

from which we obtain the expansion coefficients c_n in Eq. (2.1). Thus, the series expansion of $f(x)$ becomes

$$f(x) = \sum_{n=0}^{\infty} \frac{f^{(n)}(x_0)}{n!}(x - x_0)^n. \tag{2.2}$$

The right-hand side of Eq. (2.2) is called the *Taylor series* for the function $f(x)$. Any function $f(x)$ that is analytic in the interval $a < x < b$ may be represented by its Taylor series expanded about any point x_0 in the interval. The functions encountered in this book will be analytic in this sense.[4]

---

[4]A more general definition applicable to functions of a complex variable will be given in Chapter 7.

For example, the function

$$f(z) = (1 - z)^s$$

is analytic for any real s with z in the interval $-1 < z < 1$. We see that its derivatives are

$$f'(z) = -s(1 - z)^{s-1}$$
$$f''(z) = (-1)^2 s(s - 1)(1 - z)^{s-2}$$
$$\vdots \quad = \quad \vdots$$
$$f^{(k)}(z) = (-1)^k s(s - 1)(s - 2) \cdots (s - k + 1)(1 - z)^{s-k},$$

from which we can construct the Taylor series,

$$(1 - z)^s = \sum_{k=0}^{\infty} \frac{[f^{(k)}(z)]_{z=0}}{k!} z^k$$

$$= \sum_{k=0}^{\infty} \frac{(-1)^k s(s - 1) \cdots (s - k + 1)}{k!} z^k.$$

A more compact expression for this series may be obtained by representing the product by the *Pochhammer symbol* $(a)_n$ defined by

$$(a)_0 = 1,$$
$$(a)_n = a(a + 1)(a + 2)(a + 3) \cdots (a + n - 1), \qquad n = 1, 2, 3, \ldots. \quad (2.3)$$

Some properties of this quantity are given in Table 2.1. With this definition we write

$$(1 - z)^s = \sum_{k=0}^{\infty} \frac{(-s)_k}{k!} z^k, \tag{2.4}$$

where we have used from Table 2.1 the identity

$$(-s)_k = (-1)^k (s - k + 1)_k. \tag{2.5}$$

With $s = -1$, this reduces to the *geometric series*

$$\frac{1}{1 - z} = \sum_{k=0}^{\infty} z^k, \tag{2.6}$$

which converges for $|z| < 1$.

As a second example we construct the Taylor series for the function

$$f(x) = \log(1 + x)$$

about $x = 0$. The nth derivative is

$$f^{(n)}(x) = \frac{(-1)^{n-1}(n - 1)!}{(1 + x)^n}.$$

Clearly,

$$f(0) = 0,$$

$$f^{(n)}(x)\big|_{x=0} = (-1)^{n-1}(n-1)!, \qquad n \geq 1.$$

Thus,

$$\log(1 + x) = \sum_{n=1}^{\infty} \frac{(-1)^{n-1}}{n} x^n, \qquad |x| < 1. \tag{2.7}$$

TABLE 2.1. Identities involving Pochhammer symbols

1. $n! = (n - m)!(n - m + 1)_m$

2. $(c - m + 1)_m = (-1)^m(-c)_m$

3. $(n + m)! = n!(n + 1)_m$

4. $n! = m!(m + 1)_{n-m}$

5. $(2n - 2m)! = 2^{2n-2m}(n - m)!\left(\frac{1}{2}\right)_{n-m}$

6. $(c)_{n+m} = (c)_n(c + n)_m$

7. $(c)_n = (-1)^m(c)_{n-m}(-c - n + 1)_m$

8. $(c)_n = (-1)^{n-m}(c)_m(-c - n + 1)_{n-m}$

9. $(-n)_{m-k} = (-n)_{m-n}(m - 2n)_{n-k}$

2.2.1 SERIES EXPANSION WITH REMAINDER

We denote the difference between an analytic function $f(x)$ and the first n terms of its Taylor series by R_n. That is,

$$f(x) = \sum_{k=0}^{n-1} \frac{f^{(k)}(a)}{k!}(x - a)^k + R_n(x). \tag{2.8}$$

If $f(x)$ is analytic everywhere in the interval $a \leq x \leq b$, then

$$f(b) = \sum_{k=0}^{n-1} \frac{f^{(k)}(a)}{k!}(b - a)^k + \lambda_n(b - a)^n, \tag{2.9}$$

where we have written the remainder term $R_n(b)$ in the form

$$R_n(b) = \lambda_n(b-a)^n.$$

The number λ_n depends on b.

Let us find an expression for λ_n. We define a function $F(x)$ by

$$F(x) = -f(b) + \sum_{k=0}^{n-1} \frac{f^{(k)}(x)}{k!}(b-x)^k + \lambda_n(b-x)^n. \qquad (2.10)$$

Clearly,

$$F(b) = 0.$$

Also from Eq. (2.9) we see that

$$F(a) = 0.$$

Now, $F(x)$, along with the first $n-1$ derivatives of $f(x)$, is continuous everywhere in the interval $a \le x \le b$. So we have

$$\frac{dF(x)}{dx} = \frac{f^{(n)}(x)}{(n-1)!}(b-x)^{n-1} - n\lambda_n(b-x)^{n-1}. \qquad (2.11)$$

Since $F(a) = F(b) = 0$, there must be at least one value of x, say x_1, in the interval $a < x_1 < b$ such that $F'(x_1) = 0$.[5] In geometrical terms, if the curve representing $F(x)$ touches the x axis at $x = a$ and $x = b$, then at least at one point in the interval the tangent to the curve must be parallel to the x axis. This is illustrated in Figure 2.2. Thus, with $F'(x_1) = 0$ it follows from Eq. (2.11) that

$$\lambda_n = \frac{f^{(n)}(x_1)}{n!}, \qquad a < x_1 < b,$$

where x_1 depends on b.

Now let b in Eq. (2.9) be *any* point $x > a$. Then,

$$f(x) = \sum_{k=0}^{n-1} \frac{f^{(k)}(a)}{k!}(x-a)^k + \frac{f^{(n)}(x_1)}{n!}(x-a)^n \qquad (2.12)$$

with $a < x_1 < x$. By interchanging the roles of a and b above we find that Eq. (2.12) also holds for $x < a$ provided $x < x_1 < a$. In either case we see that the remainder in Eq. (2.8) is

$$R_n = \frac{f^{(n)}(x_1)}{n!}(x-a)^n. \qquad (2.13)$$

Often, the remainder can be shown to have an absolute value less than some known value. This is particularly useful in approximating infinite series by a finite number of terms.

---

[5]This is known as *Rolle's theorem*.

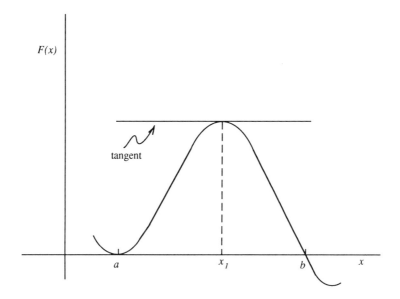

Figure 2.2

An integral expression for the remainder R_n is obtained in the following way.[6] By differentation, we have

$$\frac{d}{dt}\left[\sum_{k=0}^{n-1}\frac{(1-t)^k}{k!}(x-a)^k\frac{d^k}{d\rho^k}f(\rho)\Bigg|_{\rho=a+t(x-a)}\right]$$

$$=\frac{(1-t)^{n-1}}{(n-1)!}\frac{d^n}{dt^n}f(a+t(x-a)).$$

On integrating both sides over the interval $0 \leq t \leq 1$ we get

$$f(x)-\sum_{k=0}^{n-1}\frac{f^{(k)}(a)}{k!}(x-a)^k = \frac{1}{(n-1)!}\int_0^1(1-t)^{n-1}\left[\frac{d^n}{dt^n}f(a+t(x-a))\right]dt.$$

From this equation we see that the remainder in Eq. (2.8) is given by

$$R_n = \frac{1}{(n-1)!}\int_0^1(1-t)^{n-1}\left[\frac{d^n}{dt^n}f(a+t(x-a))\right]dt. \qquad (2.14)$$

---

[6]E.T. Whittaker and G.N. Watson, *A Course in Modern Analysis,* Cambridge University Press, Cambridge, 1927, p. 95.

As an example the function $(1-x)^b$ can be written as

$$(1-x)^b = \sum_{k=0}^{n-1} \frac{(-b)_k}{k!} x^k + R_n \qquad (2.15)$$

with the remainder given by Eq. (2.13),

$$R_n = \frac{(-b)_n}{n!} (1-x_1)^{b-n} x^n. \qquad (2.16)$$

According to Eq. (2.14) the remainder in Eq. (2.15) may be expressed as

$$R_n = \frac{x^n}{(n-1)!} (-b)_n \int_0^1 (1-t)^{n-1} (1-tx)^{b-n} dt. \qquad (2.17)$$

With the change of variable $1 - tx = (1-x)/(1-u)$ Eq. (2.17) becomes

$$R_n = \frac{(-b)_n}{(n-1)!} (1-x)^b \int_0^x u^{n-1} (1-u)^{-b-1} du. \qquad (2.18)$$

2.2.2 INTEGRATION OF INFINITE SERIES

Infinite series that converge uniformly can be integrated term by term. This property is often useful in finding series representations of certain functions. For example, if we set $z = -x^2$ in Eq. (2.6), we have

$$\frac{1}{1+x^2} = \sum_{k=0}^{\infty} (-1)^k x^{2k}, \qquad |x| < 1.$$

The integral of the left hand side is $\tan^{-1} x$. Now integrate the right hand side term by term. This gives

$$\tan^{-1} x = \sum_{k=0}^{\infty} \frac{(-1)^k}{2k+1} x^{2k+1}. \qquad (2.19)$$

In writing this result, we have chosen the constant of integration such that $\tan^{-1} x = 0$ for $x = 0$.

2.2.3 INVERSION OF SERIES

The power series

$$y(x) = \sum_{k=0}^{\infty} a_k (x-x_0)^k \qquad (2.20)$$

can be *inverted* provided $a_1 \neq 0$. By inverting the series in Eq. (2.20) we mean expand x as a Taylor series in y. Set

$$x - x_0 = \sum_{m=1}^{\infty} b_m (y-a_0)^m. \qquad (2.21)$$

By taking the series for $y - a_0$ from Eq. (2.20) and substituting into Eq. (2.21), we find that

$$x - x_0 - \sum_{m=1}^{\infty} b_m \left(\sum_{k=1}^{\infty} a_k (x - x_0)^k \right)^m = 0. \qquad (2.22)$$

Because this equation is valid for arbitrary x, coefficients of different powers of $x - x_0$ must each vanish separately. This provides us with a set of equations from which we can obtain the b_m in terms of the a_k.[7]

For simplicity, we consider the case $x_0 = 0$ and write down the first few of these equations:

$$1 - a_1 b_1 = 0$$
$$a_2 b_1 + a_1^2 b_2 = 0$$
$$a_3 b_1 + 2a_1 a_2 b_2 + a_1^3 b_3 = 0$$
$$a_4 b_1 + (2a_1 a_3 + a_2^2) b_2 + 3a_1^2 a_2 b_3 + a_1^4 b_4 = 0$$
$$a_5 b_1 + 2(a_1 a_4 + a_2 a_3) b_2 + 3(a_1 a_2^2 + a_1^2 a_3) b_3 + 4a_1^3 a_2 b_4 + a_1^5 b_5 = 0$$
$$\vdots$$

Once the coefficients b_m have been determined by this method, standard tests for convergence can be applied to the series in Eq. (2.21).

To illustrate the procedure, take $y = \tan^{-1} x$. On writing Eq. (2.19) in the form of Eq. (2.20), we see that

$$a_k = \begin{cases} \dfrac{(-1)^{(k-1)/2}}{k}, & k = \text{odd}; \\[2mm] 0, & k = \text{even}. \end{cases}$$

By following the recipe indicated above, we find that

$$x = \tan y = y + \frac{1}{3} y^3 + \frac{2}{15} y^5 + \frac{17}{315} y^7 + \cdots . \qquad (2.23)$$

We shall have occasion to use this expansion later.

2.3 Linear Second-Order Differential Equations

Now return to those special functions that appear as solutions of differential equations. Any linear, second-order, homogeneous differential equation can

---

[7]For an alternate method of evaluating the coefficients b_m in Eq. (2.21), see P.M. Morse and H. Feshbach, *Methods of Theoretical Physics*, McGraw-Hill, New York, 1953, p. 411.

be written in the form

$$\frac{d^2}{dz^2} u(z) + P(z)\frac{d}{dz} u(z) + Q(z)u(z) = 0. \tag{2.24}$$

The problem consists in finding u as an explicit function of z.

Equation (2.24) can be written as

$$u''(z) = f(z, u, u'). \tag{2.25}$$

In this form it is clear that if u and u' are given at any point z_0, then the differential equation yields u'' at z_0.

On differentiating Eq. (2.25), we get the third derivative of $u(z)$,

$$u^{(3)}(z) = \frac{df}{dz} = f'(z, u, u'),$$

which can also be evaluated at z_0. Successive differentiations give all higher derivatives (which exist if $u(z)$ is analytic). By evaluating these at z_0 we can construct the Taylor series for $u(z)$,

$$u(z) = \sum_{n=0}^{\infty} \frac{u^{(n)}(z_0)}{n!} (z - z_0)^n.$$

If this series has a nonzero radius of convergence (in Figure 2.1, $R \neq 0$), then the solution exists.

2.3.1 SINGULARITIES OF A DIFFERENTIAL EQUATION

If u and u' can be assigned arbitrary values at $z = z_0$, then we say that the point z_0 is an *ordinary point* of the differential equation, Eq. (2.24). In physical terms, this means, for example, that given some time t_0 in the oscillator problem in Section 1.3 we could choose the position $x(t_0) = x_0$ and the velocity $x'(t_0) = v_0$ both arbitrarily at that time.

If it is *not* possible to choose u and u' arbitrarily at $z = z_0$, then z_0 is a *singular point* of the differential equation or, equivalently, the differential equation has a *singularity* at z_0.[8] As an example consider the equation

$$z^2 u''(z) + azu'(z) + bu(z) = 0,$$

where a and b are constants. On setting $z = 0$ we see that if $u(0)$ has any value other than zero, either $u'(0)$ or $u''(0)$ must be infinite and the Taylor series for $u(z)$ cannot be constructed about $z = 0$.

---

[8]The definition is often given in the following form: If both $P(z)$ and $Q(z)$ are analytic at z_0, then z_0 is an ordinary point of Eq. (2.24). Otherwise, the point is singular.

2.3.2   SINGULARITIES OF A FUNCTION

It is also true that *functions,* as well as differential equations, may have singularities. There are three types: branch points, poles, and essential singularities. The *branch points* are singularities of certain multivalued functions that are not of concern here.[9] The other two are defined below. If the function $f(z)$ is analytic at z_0, we say that it is *regular* at that point. If $f(z)$ is *not* analytic at z_0, then it is *irregular* there.

If the function $f(z)$ is irregular at z_0, but the function

$$F(z) = (z - z_0)^n f(z), \qquad n = \text{positive integer},$$

is analytic at z_0, then z_0 is a *pole* of $f(z)$ of order n.

If $f(z)$ is irregular at z_0 and has neither a pole nor a branch point at that point, then $f(z)$ has an *essential singularity* at z_0. For example, if $z - z_0$ is real and positive, then for the function

$$f(z) = e^{1/(z-z_0)}$$

we have $f(z) \xrightarrow[z \to z_0]{} \infty$. But, no matter what we choose for n in $F(z)$ we still find that

$$\lim_{z \to z_0} F(z) = \lim_{z \to z_0} (z - z_0)^n e^{1/(z-z_0)} \to \infty,$$

and the singularity does not go away.

2.3.3   REGULAR AND IRREGULAR SINGULARITIES OF A DIFFERENTIAL EQUATION

We classify singular points of Eq. (2.24) as *regular* or *irregular* in the following way. If $P(z)$ or $Q(z)$ has a singularity (not a branch point) at z_0 so that $u''(z_0)$ cannot be obtained for constructing the Taylor series of $u(z)$, then the differential equation has a *regular* singularity if and only if both $(z - z_0)P(z)$ and $(z - z_0)^2 Q(z)$ are analytic at z_0. Otherwise, the singularity at $z = z_0$ is *irregular.*

2.4   The Hypergeometric Equation

All of the differential equations we encounter in this book have at most three singularities. For these equations, the functions $P(z)$ and $Q(z)$ in Eq. (2.24) are *rational.*[10] By an appropriate change of variable, these singularities

---

[9]See Chapter 7.

[10]A function $f(z)$ is rational if it can be written as a ratio of two polynomials. If $f(z) = p(z)/q(z)$, where $p(z)$ and $q(z)$ are polynomials in z, then $f(z)$ is a rational

may be transformed to three points $(0, 1, \infty)$, in which case the differential equation takes the form

$$z(1 - z)\frac{d^2u}{dz^2} + [c - (a + b + 1)z]\frac{du}{dz} - abu = 0. \qquad (2.26)$$

This is Gauss's *hypergeometric* equation. The parameters a, b, and c are independent of z and, in general, may be complex.[11]

Since all of the differential equations we consider can be related to Eq. (2.26) (or a certain limiting case of it), it is sufficient to solve this equation once and then express all of the solutions of the equations we discuss subsequently in terms of the solutions to it.

In applying the series method we assume a solution of the form

$$u(z) = \sum_{n=0}^{\infty} a_n z^{n+s} \qquad (2.27)$$

and substitute this series into Eq. (2.26) to obtain

$$(z - z^2)\sum_{n=0}^{\infty} a_n(n + s)(n + s - 1)z^{n+s-2}$$

$$+ [c - (a + b + 1)z]\sum_{n=0}^{\infty} a_n(n + s)z^{n+s-1} - ab\sum_{n=0}^{\infty} a_n z^{n+s} = 0.$$

Now let us collect terms with powers of z which look the same. Then

$$\sum_{n=0}^{\infty} a_n(n+s)(n+s+c-1)z^{n+s-1} - \sum_{n=0}^{\infty} a_n[(n+s)(n+s+a+b)+ab]z^{n+s} = 0.$$

By redefining the summation index in the first sum in this equation and combining coefficients of the same power of z, we can write

$$a_0 s(s + c - 1)z^{s-1} + \sum_{n=0}^{\infty}\{(n + s + 1)(n + s + c)a_{n+1}$$

$$- [(n + s + a + b)(n + s) + ab]a_n\}z^{n+s} = 0.$$

Since this equation holds for arbitrary values of z, the coefficient of each power of z must vanish separately. This leads to the relations,

$$s(s + c - 1)a_0 = 0 \qquad (2.28)$$

---

function of z. For further discussion of the singularities of the linear differential equations in mathematical physics, see E.T. Whittaker and G.N. Watson, *op. cit.*, p. 203.

[11] See Chapter 7 for a discussion of complex numbers.

and

$$a_{n+1} = \frac{(n+s)(n+s+a+b)+ab}{(n+s+1)(n+s+c)} a_n. \tag{2.29}$$

From Eq. (2.29) we see that if $a_0 = 0$, then $a_n = 0$ for *all* values of n and $u(z) = 0$ is the only solution. Therefore, to obtain a nontrivial solution we must have $a_0 \neq 0$. Then, according to Eq. (2.28) either $s = 0$ or $s = 1 - c$. Let us consider the case for $s = 0$ first. We set $s = 0$ in Eq. (2.29) and rearrange terms in the numerator to get

$$a_{n+1} = \frac{(n+a)(n+b)}{(n+1)(n+c)} a_n. \tag{2.30}$$

The first few of these coefficients are

$$a_1 = \frac{ab}{c} a_0,$$

$$a_2 = \frac{(a+1)(b+1)}{2(c+1)} a_1 = \frac{a(a+1)b(b+1)}{1 \cdot 2c(c+1)} a_0,$$

$$a_3 = \frac{(a+2)(b+2)}{3(c+2)} a_2 = \frac{a(a+1)(a+2)b(b+1)(b+2)}{1 \cdot 2 \cdot 3c(c+1)(c+2)} a_0.$$

If we continue the pattern, we see that the nth coefficient is

$$a_n = \frac{a(a+1)(a+2)\cdots(a+n-1)b(b+1)(b+2)\cdots(b+n-1)}{n!c(c+1)(c+2)\cdots(c+n-1)} a_0.$$

So, as a solution to Eq. (2.26) we obtain

$$u(z) = a_0 \left[1 + \sum_{n=1}^{\infty} \frac{a(a+1)\cdots(a+n-1)b(b+1)\cdots(b+n-1)}{n!c(c+1)\cdots(c+n-1)} z^n \right]. \tag{2.31}$$

By using Pochhammer symbols we can write for the expression in square brackets in Eq. (2.31)

$$F(a,b;c;z) = \sum_{n=0}^{\infty} \frac{(a)_n (b)_n}{n!(c)_n} z^n. \tag{2.32}$$

The function $F(a, b; c; z)$ defined by this series is the *hypergeometric function*.[12]

---

[12] Or *Gauss's hypergeometric function*. See A. Erdélyi, Ed., *Higher Transcendental Functions*, McGraw-Hill, New York, 1955, Vol. 1, p. 56 ff.

2.4.1 EXAMPLES

If we set $a = 1$ and $c = b$ in Eq. (2.32), we get

$$F(1, b; b; z) = \sum_{n=0}^{\infty} z^n, \tag{2.33}$$

which is the geometric series of Eq. (2.6). Thus, the hypergeometric series is a generalization of the geometric series. Other familiar expressions may also be written as hypergeometric functions. For example, we see from Eq. (2.4) that

$$(1 - z)^s = F(-s, b; b; z). \tag{2.34}$$

2.4.2 LINEARLY INDEPENDENT SOLUTIONS

From inspection of Eq. (2.31) we can see that

a. If a or b is a negative integer or zero, then the series terminates (i.e., it is a polynomial).

b. If c is a negative integer or zero, then the series requires special handling. We shall consider this case later.

The hypergeometric equation (Eq. 2.26) is a homogeneous, linear, second-order differential equation. The general solution consists of a linear combination of two linearly independent solutions $u_1(z)$ and $u_2(z)$,

$$u(z) = c_1 u_1(z) + c_2 u_2(z). \tag{2.35}$$

By *linear independence* we mean that in Eq. (2.35) $u(z) = 0$ for all z if and only if $c_1 = c_2 = 0$.[13]

We have one solution

$$u_1(z) = F(a, b; c; z),$$

which we obtained by taking $s = 0$ in Eq. (2.29). To get a second solution we let $s = -c + 1$. With some rearrangement in the numerator and denominator of Eq. (2.29), we get

$$a_{n+1} = \frac{[n + (2 - c) - 1][n + (2 - c) - 1 + a + b] + ab}{(n + 1)(n + (2 - c))} a_n.$$

---

[13]This definition is easily generalized to a sum of n terms. Linear independence of the functions $u_k(z)$ in the equation

$$\sum_{k=1}^{n} c_k u_k(z) = 0,$$

where z is arbitrary, requires that $c_k = 0$ for all k.

If we set $c' = 2 - c$ and rearrange the terms in the numerator, we have

$$a_{n+1} = \frac{(n + c' - 1 + a)(n + c' - 1 + b)}{(n+1)(n+c')} a_n.$$

Here we see that the denominator has the same form as that in Eq. (2.30). We can transform the numerator by setting

$$a' = 1 + a - c \quad \text{and} \quad b' = 1 + b - c.$$

With these definitions we have

$$a_{n+1} = \frac{(n + a')(n + b')}{(n+1)(n+c')} a_n.$$

This expression has exactly the same form as Eq. (2.30), which generated the coefficients a_n in Eq. (2.27) for $s = 0$. So for $s = 1-c$ our second solution is

$$u_2(z) = z^{1-c} F(1 + a - c, 1 + b - c; 2 - c; z).$$

The general solution of Eq. (2.26) is then

$$u(z) = AF(a, b; c; z) + Bz^{1-c}F(1 + a - c, 1 + b - c; 2 - c; z), \quad (2.36)$$

where A and B are the arbitrary constants of integration to be determined by the boundary conditions.

2.4.3 If c Is an Integer

Note that if $c = 1$, then $u_2(z)$ is *not* a new solution. In fact, if c is any integer at all, $u_1(z)$ and $u_2(z)$ as defined above may not represent two linearly independent solutions. To see that this is so, let us examine the series representations of the two solutions,

$$u_1(z) = \sum_{k=0}^{\infty} \frac{(a)_k (b)_k}{k!(c)_k} z^k, \quad (2.37a)$$

$$u_2(z) = z^{1-c} \sum_{k=0}^{\infty} \frac{(a + 1 - c)_k (b + 1 - c)_k}{k!(2 - c)_k} z^k. \quad (2.37b)$$

We now consider two cases where c is an integer ($\neq 1$).

$c = n \geq 2$. The factor in the denominator of each term in the sum in $u_2(z)$ can be written

$$(2 - c)_k = (2 - n)_k = (2 - n)(3 - n) \cdots (-1) \cdot 0 \cdot (-n + k + 1)!,$$

which contains a factor zero for all terms with $k \geq n - 1$. Thus, we see that the denominator in each of the higher order terms vanishes and, in general,

$u_2(z)$ cannot represent a solution of the hypergeometric equation for such values of c.[14]

$c = -n \leq 0$. For $u_1(z)$ we can write the factor in the denominator of each term in the sum as

$$(c)_k = (-n)_k = (-n)(-n+1)\cdots(-1)\cdot 0 \cdot (-n+k-1)!,$$

which contains a zero in the product for all terms with $k \geq n+1$. Therefore, the denominator in each of the higher order terms goes to zero and $u_1(z)$ is not a solution.

If c is an integer and a (or b) is *also* an integer, then it may be possible to have solutions given by both series. For example, suppose we have $c = -n \leq 0$ and $a = -m \leq 0$ with $m < n$. Then the ratio of the corresponding factors in the kth term of the series for $u_1(z)$ is

$$\frac{(a)_k}{(c)_k} = \frac{(m-k+1)_{n-m}}{(m+1)_{n-m}}. \tag{2.38}$$

Clearly, the denominator on the right hand side does not vanish for any k while the numerator is zero for all k in the range $m + 1 \leq k \leq n$. Note that if $m > n$, then the roles of numerator and denominator in this ratio would be interchanged and again we would not have a solution. Similar conditions may be obtained for $u_2(z)$ with $c = n \geq 2$.[15]

2.5   The Simple Pendulum

For an application of the hypergeometric function consider the exact solution of the simple pendulum problem. A simple pendulum consists of a point mass m attached to one end of a massless rod of length L. The other end of the rod is fixed at a point such that the system can swing freely under the influence of gravity. See Figure 2.3a.

The two forces exerted on the mass are the weight mg and the tension T in the rod as shown in Figure 2.3b. On resolving the equation of motion (Newton's second law) into two components we have

$$\text{centripetal:} \quad T - mg\cos\theta = \frac{mv^2}{L},$$

$$\text{tangential:} \quad -mg\sin\theta = m\frac{d^2}{dt^2}(L\theta).$$

---

[14] For linearly independent solutions in this case see H. Jeffreys and B.S. Jeffreys, *op. cit.*, p. 482.
[15] See Exercise 2.6.

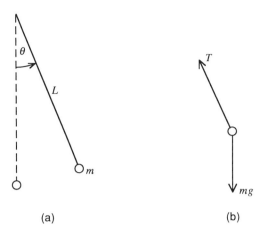

(a)   (b)

Figure 2.3

By rearranging the tangential equation we get

$$\frac{d^2\theta}{dt^2} + \frac{g}{L}\sin\theta = 0. \tag{2.39}$$

The usual approximation for small-amplitude oscillations is to take the leading term in the series expansion of $\sin\theta$ in which case the motion described is that of a simple harmonic oscillator (cf. Eq. 1.6).

To obtain an exact solution to Eq. (2.39) we start by multiplying by $d\theta/dt$. This yields

$$\frac{d}{dt}\left[\frac{1}{2}\left(\frac{d\theta}{dt}\right)^2 - \frac{g}{L}\cos\theta\right] = 0,$$

which means that the quantity in square brackets is a constant.[16] To get a value for the constant, assume that we release the pendulum from rest when the angle θ has the value θ_0. This leads to

$$\frac{d\theta}{dt} = \sqrt{\frac{2g}{L}(\cos\theta - \cos\theta_0)}.$$

---

[16]In fact, if we multiply the quantity by mL^2, we see that the first term is the kinetic energy and the second is the gravitational potential energy. The equation then becomes a statement of energy conservation.

With the identity $\cos 2x = 1 - 2\sin^2 x$ and some rearrangement of the equation, we get

$$2\sqrt{\frac{g}{L}} \int dt = \int \frac{d\theta}{[k^2 - \sin^2(\theta/2)]^{1/2}}.$$

In the integral, $k = \sin(\theta_0/2)$. Because $\theta \le \theta_0$, we can make the change of variable $\sin\phi = k^{-1}\sin(\theta/2)$ to obtain

$$\sqrt{\frac{g}{L}} \int dt = \int \frac{d\phi}{\sqrt{1 - k^2 \sin^2 \phi}}.$$

The integral on the right is an *elliptic integral of the first kind* denoted by

$$F(k, \phi) = \int_0^\phi \frac{d\phi'}{\sqrt{1 - k^2 \sin^2 \phi'}}.$$

The time required for the angle to change from 0 to θ_0 is one quarter of the period T. In this time ϕ goes from 0 to $\pi/2$. With these limits we have for the integral

$$F(k) = \int_0^{\pi/2} \frac{d\phi}{\sqrt{1 - k^2 \sin^2 \phi}} \tag{2.40}$$

which is called the *complete* elliptic integral of the first kind.

To evaluate $F(k)$ use Eq. (2.4) to expand the integrand

$$[1 - k^2 \sin^2 \phi]^{-1/2} = \sum_{n=0}^\infty \frac{\left(\frac{1}{2}\right)_n}{n!} k^{2n} \sin^{2n} \phi.$$

On substituting this expansion into Eq. (2.40) we see that we must evaluate an integral of the type

$$\int \sin^p x \, dx = \int \sin^{2n} \phi \, d\phi.$$

Integrate by parts to obtain

$$\int \sin^p x \, dx = -\sin^{p-1} x \cos x + (p-1) \int \sin^{p-1} x \cos^2 x \, dx$$

$$= -\sin^{p-1} x \cos x + (p-1) \int \sin^{p-2} x \, dx - (p-1) \int \sin^p x \, dx.$$

On solving for $\int \sin^p x \, dx$, we arrive at the formula

$$\int \sin^p x \, dx = -\frac{\sin^{p-1} x \cos x}{p} + \frac{p-1}{p} \int \sin^{p-2} x \, dx.$$

By applying the formula again to the integral on the right we have

$$\int \sin^p x \, dx = -\frac{\sin^{p-1} x \cos x}{p} - \frac{p-1}{p} \frac{\sin^{p-3} x \cos x}{(p-2)}$$
$$+ \frac{(p-1)(p-3)}{p(p-2)} \int \sin^{p-4} x \, dx.$$

Repeated applications of the formula will always result in a sum of terms, each involving a product of $\cos x$ with some power of $\sin x$, plus an integral of some power of $\sin x$. If we take the range of integration to be from 0 to $\pi/2$, then $\cos x$ will vanish at one limit and $\sin x$ will be zero at the other. Only the term with the integral will survive. The result is

$$\int_0^{\pi/2} \sin^p x \, dx$$

$$= \begin{cases} \dfrac{(p-1)(p-3)\cdots(p-(p-1))}{p(p-2)\cdots(p-(p-2))} \displaystyle\int_0^{\pi/2} dx, & p = 0, 2, 4, \ldots ; \\[3ex] \dfrac{(p-1)(p-3)\cdots(p-(p-2))}{p(p-2)\cdots(p-(p-3))} \displaystyle\int_0^{\pi/2} \sin x \, dx, & p = 1, 3, 5, \ldots . \end{cases}$$

The integrals on the right can be evaluated directly to yield $\pi/2$ and 1, respectively. The more compact Pochhammer notation for the products yields[17]

$$\int_0^{\pi/2} \sin^p x \, dx = \begin{cases} \dfrac{\pi}{2} \dfrac{\left(\frac{1}{2}\right)_{p/2}}{\left(\frac{p}{2}\right)!}, & p = 0, 2, 4, \ldots ; \\[3ex] \dfrac{((p-1)/2)!}{(3/2)_{(p-1)/2}}, & p = 1, 3, 5, \ldots . \end{cases}$$

If the result for even p is inserted into the expansion of the integrand in Eq. (2.40), the complete elliptic integral is

$$F(k) = \frac{\pi}{2} \sum_{n=0}^{\infty} \frac{\left(\frac{1}{2}\right)_n \left(\frac{1}{2}\right)_n}{n!(1)_n} k^{2n}$$
$$= \frac{\pi}{2} F\left(\tfrac{1}{2}, \tfrac{1}{2}; 1; k^2\right).$$

Thus, the period of oscillation of a simple pendulum is given by

$$T = 2\pi \sqrt{\frac{L}{g}} F\left(\tfrac{1}{2}, \tfrac{1}{2}; 1; \sin^2\left(\frac{\theta_0}{2}\right)\right).$$

---

[17] These results also follow from the properties of the gamma function. See Exercise 2.21.

Clearly, the period *does* depend on the amplitude of the oscillation. However, if the amplitude is small we can use only the leading term in the hypergeometric series. Then, the period reduces to the usual small-angle approximation

$$T = 2\pi(L/g)^{1/2}, \qquad \theta_0 << 1.$$

The elliptic integral of the second kind is defined by

$$E(k, \phi) = \int_0^\phi \sqrt{1 - k^2 \sin^2 \phi'} \, d\phi'.$$

The complete elliptic integral of the second kind is related to the hypergeometric function by

$$E(k) = \int_0^{\pi/2} \sqrt{1 - k^2 \sin^2 \phi} \, d\phi = \frac{\pi}{2} F\left(-\tfrac{1}{2}, \tfrac{1}{2}; 1; k^2\right). \tag{2.41}$$

2.6   The Generalized Hypergeometric Function

The hypergeometric function defined in Eq. (2.32) can be generalized in an obvious way

$$_pF_q(a_1, a_2, \ldots, a_p; b_1, b_2, \ldots, b_q; z)$$

$$= \sum_{n=0}^\infty \frac{(a_1)_n (a_2)_n \cdots (a_p)_n}{n!(b_1)_n (b_2)_n \cdots (b_q)_n} z^n, \tag{2.42}$$

where p is the number of numerator parameters a_i and q is the number of denominator parameters b_j. We shall frequently make use of this notation, which is due to Pochhammer and Barnes.[18] With this notation the hypergeometric function of Eq. (2.32) is $_2F_1(a, b; c; z)$.

2.7   Vandermonde's Theorem

A powerful relation that is useful in manipulating sums involving Pochhammer symbols is

$$\sum_{m=0}^n \frac{(a)_m}{m!} \frac{(b)_{n-m}}{(n-m)!} = \frac{(a+b)_n}{n!} \tag{2.43}$$

where a and b are numbers[19] independent of the summation index m. This relation is known as *Vandermonde's theorem*. To prove it we express s in

---

[18] See G.N. Watson, *A Treatise on the Theory of Bessel Functions*, Cambridge University Press, Cambridge, 1922, p. 100.

[19] These numbers may be *complex*. See Chapter 7.

Eq. (2.4) as the sum of two numbers a and b and set $z = -x$ to obtain

$$(1 + x)^{a+b} = \sum_{n=0}^{\infty} \frac{(a + b - n + 1)_n}{n!} x^n. \tag{2.44}$$

This function can also be written as a product

$$(1 + x)^{a+b} = (1 + x)^a (1 + x)^b = \sum_{k=0}^{\infty} \frac{(a - k + 1)_k}{k!} x^k \sum_{j=0}^{\infty} \frac{(b - j + 1)_j}{j!} x^j. \tag{2.45}$$

The product of any two infinite series may be rewritten as

$$\sum_{k=0}^{\infty} S_k \sum_{j=0}^{\infty} T_j = S_0 T_0 + (S_0 T_1 + S_1 T_0) + (S_0 T_2 + S_1 T_1 + S_2 T_0) + \cdots$$

$$= \sum_{n=0}^{\infty} \sum_{m=0}^{n} S_m T_{n-m}.$$

Such a product constructed in this way is called the *Cauchy product*. Clearly, the Cauchy product of two series is

$$\sum_{k=0}^{\infty} S_k \sum_{j=0}^{\infty} T_j = \sum_{n=0}^{\infty} U_n, \tag{2.46}$$

where

$$U_n = \sum_{m=0}^{n} S_m T_{n-m}.$$

If $\sum_{k=0}^{\infty} S_k$ and $\sum_{j=0}^{\infty} T_j$ are absolutely convergent, then $\sum_{n=0}^{\infty} U_n$ is absolutely convergent.[20]

We see that the Cauchy product of the two series in Eq. (2.45) is

$$(1 + x)^{a+b} = \sum_{n=0}^{\infty} \sum_{m=0}^{n} \frac{(a - m + 1)_m}{m!} \frac{(b - n + m + 1)_{n-m}}{(n - m)!} x^n. \tag{2.47}$$

Subtraction of Eq. (2.47) from Eq. (2.44) yields

$$\sum_{n=0}^{\infty} \left[\frac{(a + b - n + 1)_n}{n!} - \sum_{m=0}^{n} \frac{(a - m + 1)_m}{m!} \frac{(b - n + m + 1)_{n-m}}{(n - m)!} \right] x^n = 0.$$

Since this holds for arbitrary x in the interval $-1 < x < 1$, the terms in different powers of x are linearly independent. Therefore, the coefficient of each power of x must vanish separately, that is,

$$\sum_{m=0}^{n} \frac{(a - m + 1)_m}{m!} \frac{(b - n + m + 1)_{n-m}}{(n - m)!} = \frac{(a + b - n + 1)_n}{n!}. \tag{2.48}$$

---

[20] See E.C. Titchmarsh, *op. cit.*, p. 32.

Now use Eq. (2.5) to get

$$\sum_{m=0}^{n} \frac{(-a)_m}{m!} \frac{(-b)_{n-m}}{(n-m)!} = \frac{(-a-b)_n}{n!}.$$

On replacing $-a$ by a and $-b$ by b, Eq. (2.43) follows.

2.8   Leibniz's Theorem

As an application of Vandermonde's theorem we derive a formula for the mth derivative of the product of two analytic functions $u(x)$ and $v(x)$. We expand the functions as Taylor series about $x = 0$ and differentiate m times to obtain

$$\frac{d^m}{dx^m}[u(x)v(x)] = \frac{d^m}{dx^m}\left[\sum_{p=0}^{\infty} a_p x^p \sum_{q=0}^{\infty} b_q x^q\right]$$

$$= \sum_{p=0}^{\infty}\sum_{q=0}^{\infty} a_p b_q (-1)^m (-p-q)_m x^{p+q-m}. \tag{2.49}$$

By Vandermonde's theorem

$$(-p-q)_m = m!\sum_{k=0}^{m} \frac{(-p)_k}{k!} \frac{(-q)_{m-k}}{(m-k)!}.$$

On substituting this result into Eq. (2.49) and reordering the sums we get

$$\frac{d^m}{dx^m}[u(x)v(x)]$$

$$(-1)^m \sum_{k=0}^{m} \frac{m!}{k!(m-k)!} \sum_{p=0}^{\infty} a_p(-p)_k x^{p-k} \sum_{q=0}^{\infty} b_q(-q)_{m-k} x^{q-m+k}$$

$$= \sum_{k=0}^{\infty} \frac{m!}{k!(m-k)!} \left[\frac{d^k}{dx^k}\sum_{p=0}^{\infty} a_p x^p\right]\left[\frac{d^{m-k}}{dx^{m-k}}\sum_{q=0}^{\infty} b_q x^q\right].$$

We recognize the sums in this last expression as the Taylor expansions of $u(x)$ and $v(x)$, respectively. Thus,

$$\frac{d^m}{dx^m}[u(x)v(x)] = \sum_{k=0}^{m} \frac{(m-k+1)_k}{k!}\left[\frac{d^k}{dx^k}u(x)\right]\left[\frac{d^{m-k}}{dx^{m-k}}v(x)\right]. \tag{2.50}$$

This is *Leibniz's theorem*.[21]

---

[21]For an alternate proof that does not require $u(x)$ and $v(x)$ to be analytic at $x = 0$, but only that they be m times differentiable, see Exercise 2.28.

EXERCISES

1. For each of the following functions calculate enough of the coefficients of $(x-x_0)^n$ in the expansion of the Taylor series to establish a pattern for the general coefficient. Write out the general coefficient explicitly. Write down the explicit expansion of the function in the form of Eq. (2.2). The range of validity of the expansion is given. Use $x = \frac{1}{2}$ in each case to calculate the first few terms in the expansion to see that your result converges to the appropriate value of $f\left(\frac{1}{2}\right)$. Use enough terms to obtain agreement with the exact value to four significant digits. Note the number of terms you need in each case to get this agreement.

    a. $f(x) = 1/x \ (x_0 = 1)$ $0 < x < 2$
    b. $f(x) = \log x \ (x_0 = 1)$ $0 < x < 2$
    c. $f(x) = \sqrt{1-x} \ (x_0 = 0)$ $-1 < x < 1$

2. Carry out the indicated change of variable in Eq. (2.17) to show that the remainder in the series expansion of $(1-x)^b$ is given by

$$R_n = \frac{(-b)_n}{(n-1)!}(1-x)^b \int_0^x u^{n-1}(1-u)^{-b-1}du.$$

3. Verify that the identity

$$(1)_n = n!$$

follows from the definition Eq. (2.3).

4. Start with the definition Eq. (2.3) and verify the identities in Table 2.1.

5. Show that if z is any number and n is any nonnegative integer, then

$$\Gamma(z+n) = \Gamma(z)(z)_n.$$

6. Verify Eq. (2.38) and show that

$$(m-k+1)_{n-m} = 0, \quad \text{for} \quad m+1 \le k \le n.$$

By a similar calculation find the values of a in Eq. (2.37) for which $u_2(z)$ may be a solution to the hypergeometric equation with $c = n \ge 2$.

7. Use Eq. (2.4) to prove the *binomial theorem*,

$$(a+b)^n = \sum_{m=0}^n \frac{n!}{m!(n-m)!} a^{n-m}b^m$$

where n is a positive integer. For definiteness, assume that $a > b$.

8. Obtain the first four nonvanishing terms in the Taylor expansion of $\tan y$ about $y = 0$ by calculating the derivatives of $\tan y$ and evaluating them at $y = 0$. Compare with the result in Eq. (2.23).

9. The series expansion for $\sin^{-1} x$ is

$$\sin^{-1} x = \frac{1}{\sqrt{\pi}} \sum_{n=0}^{\infty} \frac{\Gamma(n + \frac{1}{2})}{n!(2n + 1)} x^{2n+1}.$$

Show that

$$\sin^{-1} x = x\, _2F_1 \left(\tfrac{1}{2}, \tfrac{1}{2}; \tfrac{3}{2}; x^2 \right).$$

10. By manipulating the kth term in the expansion in Eq. (2.19) show that

$$\tan^{-1} x = x\, _2F_1 \left(\tfrac{1}{2}, 1; \tfrac{3}{2}; -x^2 \right).$$

11. By comparing the series for $_2F_1(1, 1; 2; z)$ with the Taylor expansion given in Eq. (2.7) show that

$$\log(1 - z) = -z\, _2F_1(1, 1; 2; z).$$

12. Atoms arranged in a regular array in a metal crystal are separated by some distance. At the absolute zero of temperature $(T = 0)$ the potential energy of a pair of atoms displaced by x from their equilibrium separation is represented by[22]

$$V(x) = ax^2 - bx^3.$$

The parameters a and b are independent of x. At a finite temperature T the average displacement $\bar{x}$ is obtained from the Boltzmann distribution function according to

$$\bar{x} = \int_{-\infty}^{\infty} xe^{-V(x)/kT}\, dx \bigg/ \int_{-\infty}^{\infty} e^{-V(x)/kT}\, dx.$$

Show that if the atoms are harmonically bound $(b = 0)$, then the average displacement is zero. If the anharmonic term in $V(x)$ is small, but not zero, we can take the leading terms in the expansion of the exponential to obtain

$$\bar{x} = \gamma T$$

where γ is a constant called the *coefficient of linear expansion*. Write the integrals in terms of gamma functions to show that

$$\gamma = \frac{3kb}{4a^2}.$$

---

[22]See C. Kittel, *Introduction to Solid State Physics*, Wiley, New York, 1956, p. 152.

13. Find the values of the parameters a, b, and c for the function

$$_2F_1(a, b; c; z) = \frac{1}{\mu z}\left[\frac{1}{(1-z)^\mu} - 1\right]$$

where μ is a number.

14. By inspection write down the general solution of the differential equation

$$z(1-z)u''(z) + \left(\frac{5}{4} - 2z\right)u'(z) + \frac{3}{4}u(z) = 0.$$

15. Write the series solution obtained in Exercise 1.5 in the form of the generalized function of Eq. (2.42). Write your result using the notation $_pF_q(a_1, \ldots a_p; b_1, \ldots, b_q; z)$.

16. The general solution of the second-order differential equation

$$xy''(x) + \mu y'(x) - \lambda y(x) = 0$$

with constant parameters μ and λ may be written as

$$y(x) = Ay_1(x) + By_2(x),$$

where A and B are arbitrary constants of integration. Use the series method to find y_1 and y_2. Express your results in terms of Pochhammer symbols. Write your solution using the notation of the generalized function of Eq. (2.42). Discuss the restrictions to be imposed on parameters μ and λ for your result to be a general solution to the differential equation.

17. Solve the differential equation

$$xf''(x) + 2f'(x) + xf(x) = 0$$

by the series method. Write your results in terms of the generalized notation of Eq. (2.42). Note that the solution can also be written in terms of elementary functions.

18. For z in the range $0 < z < 1$ make the change of variable $y = 1 - z$ in the hypergeometric equation and show that a general solution to Eq. (2.26) is

$$u(z) = A\,_2F_1(a, b; a + b - c + 1; 1 - z)$$
$$+ B(1 - z)^{c-a-b}\,_2F_1(c - b, c - a; c - a - b + 1; 1 - z).$$

19. Begin with the expansions in Eqs. (1.3) and (1.4) and show that

$$\cos x = {}_0F_1\left(\frac{1}{2}; \frac{-x^2}{4}\right) \quad \text{and} \quad \frac{\sin x}{x} = {}_0F_1\left(\frac{3}{2}; \frac{-x^2}{4}\right).$$

20. Show that the complete elliptic integral of the second kind can be expressed in terms of the hypergeometric function as in Eq. (2.41).

21. Show that the results obtained for the integral

$$\int_0^{\pi/2} \sin^p x \, dx$$

in the text are consistent with the more general result of Exercise 1.16.

22. Start with Eq. (1.22) and make use of Eq. (2.4) to show that for $x > 0$,

$$\frac{\Gamma(x)}{\Gamma(x + \frac{1}{2})} = \frac{1}{\sqrt{\pi}} \sum_{k=0}^{\infty} \frac{(2k)!}{2^{2k} k! k!} \frac{1}{x + k}.$$

23. Write down the general solution to the hypergeometric equation

$$z(1 - z)u''(z) + \mu(1 - z)u'(z) + \mu u(z) = 0,$$

where μ is not an integer. Evaluate the two hypergeometric series in the solutions and verify that one of the solutions is a polynomial.

24. Use the properties of the Pochhammer symbol in the hypergeometric series to show that

$${}_2F_1(a, b+1; c; z) - {}_2F_1(a, b; c; z) = \frac{az}{c} {}_2F_1(a+1, b+1; c+1; z), \quad |z| < 1.$$

25. Show that

$${}_2F_1(a, c + 1; c; z) = \left[1 - \left(1 - \frac{a}{c}\right)z\right](1 - z)^{-a-1}, \quad |z| < 1.$$

26. By following the procedure leading to Eq. (2.46) show that the Cauchy product of two series can be generalized to negative indices,

$$\sum_{k=-\infty}^{\infty} A_k \sum_{j=-\infty}^{\infty} B_j = \sum_{n=-\infty}^{\infty} \sum_{m=-\infty}^{\infty} A_m B_{n-m}.$$

27. Use Leibniz's theorem to calculate the mth derivative of the product

$$x^a e^x.$$

Write out your result for arbitrary a then consider the special case where a is an integer less than m. How must your general result be modified for this special case?

28. Prove Leibniz's theorem (Eq. 2.50) by mathematical induction. This will involve some manipulations of Pochhammer symbols. Note that this proof does not require that $u(x)$ and $v(x)$ be analytic, but only that each may be differentiated m times.

3

The Confluent Hypergeometric Function

Many of the special functions of mathematical physics can be expressed in terms of specific forms of the *confluent hypergeometric function.* As its name suggests, this function is related to the hypergeometric function of Chapter 2. Let us see how.

3.1   The Confluent Hypergeometric Equation

Our motive here is to derive the confluent hypergeometric equation from the hypergeometric equation in a manner that emphasizes the connection between the two equations and their solutions.[1] The hypergeometric series (Eq. 2.32) can be written

$$F(a, b; c; z) = \sum_{n=0}^{\infty} \frac{(a)_n b(b+1) \cdots (b+n-1)}{n!(c)_n} z^n$$

$$= \sum_{n=0}^{\infty} \frac{(a)_n (1)\left(1 + \frac{1}{b}\right) \cdots \left(1 + \frac{n-1}{b}\right)}{n!(c)_n} x^n,$$

where we have removed a factor b^n from $(b)_n$ and defined $x = bz$. In this last expression, if we take the limit as b tends to infinity, we obtain the function

$$M(a, c; x) = \sum_{n=0}^{\infty} \frac{(a)_n}{n!(c)_n} x^n.$$

We see from the definition of the generalized hypergeometric function (Eq. 2.42) that

$$M(a, c; x) = {}_1F_1(a; c; x).$$

How is the hypergeometric equation itself affected in this limit? Make the substitution $x = bz$ in Eq. (2.26) and obtain

$$x \left(1 - \frac{x}{b}\right) \frac{d^2 u}{dx^2} + \left[c - \left(\frac{a+1}{b} + 1\right) x\right] \frac{du}{dx} - au = 0.$$

---

[1] For a more rigorous treatment see E.T. Whittaker and G.N. Watson, *op. cit.*, p. 337.

Now take the limit $b \rightarrow \infty$, which yields the *confluent hypergeometric equation*,[2]

$$x \frac{d^2 u}{dx^2} + (c - x) \frac{du}{dx} - au = 0. \tag{3.1}$$

This equation has singularities at $x = 0$ and $x = \infty$. What kind are they? First, we write Eq. (3.1) in the standard form of Eq. (2.24) with

$$P(x) = \frac{c}{x} - 1 \quad \text{and} \quad Q(x) = -\frac{a}{x}.$$

We note that

$$xP(x) = c - x \quad \text{and} \quad x^2 Q(x) = -ax$$

are both analytic at $x = 0$. Thus, the singularity at $x = 0$ is regular.

To examine the behavior at $x = \infty$ we change the variable to $y = 1/x$ and consider the singularity at $y = 0$. With this change Eq. (3.1) can be written as

$$\frac{d^2 u}{dy^2} + \frac{y + (2 - c)y^2}{y^3} \frac{du}{dy} - \frac{a}{y^3} u = 0.$$

From a comparison with Eq. (2.24) we see that

$$yP(y) = (1 + (2 - c)y)/y$$

and

$$y^2 Q(y) = -a/y,$$

neither of which is analytic at $y = 0$.

Two things have happened in arriving at Eq. (3.1) from Eq. (2.26) in the limit considered:

a. A merging (*confluence*) of the singularities of Eq. (2.26) at $z = 1$ and $z = \infty$ has occurred.

b. The singularity at infinity is now *irregular*.

Now that we have seen how the confluent hypergeometric equation is related to the hypergeometric equation and how it gets its name, let us go back to Eq. (3.1) and apply the series method of Chapter 2 to obtain a general solution. On substituting the series

$$u(x) = \sum_{k=0}^{\infty} a_k x^{k+s}$$

into Eq. (3.1) and requiring that the resulting equation hold for arbitrary x we obtain the relations

$$s(s - 1 + c)a_0 = 0$$

---

[2] At this point we have *not* shown that $u(x) = {}_1F_1(a; c; x)$ is a solution to Eq. (3.1).

and

$$a_{k+1} = \frac{k+s+a}{(k+s+1)(k+s+c)} a_k.$$

For a nontrivial solution clearly $a_0 \neq 0$. Thus, either $s = 0$ or $s = 1 - c$, leading to

$$u_1(x) = a_0 \sum_{k=0}^{\infty} \frac{(a)_k}{k!(c)_k} x^k = a_0 \, {}_1F_1(a; c; x), \quad s = 0,$$

and

$$u_2(x) = a_0 \sum_{k=0}^{\infty} \frac{(1+a-c)_k}{k!(2-c)_k} x^{k+1-c} = a_0 x^{1-c} \, {}_1F_1(1+a-c; 2-c; x), \quad s = 1-c,$$

where we have used the generalized notation of Eq. (2.42).

If c is not an integer, then the general solution of Eq. (3.1) is

$$u(x) = A \, {}_1F_1(a; c; x) + B x^{1-c} \, {}_1F_1(1 + a - c; 2 - c; x). \tag{3.2}$$

3.2   One-Dimensional Harmonic Oscillator

As an application we treat the familiar problem of a quantum harmonic oscillator in one dimension. First, consider the time-dependent Schrödinger equation for a particle of mass m with a potential energy $V(x)$, which depends on position only,

$$\left[\frac{-\hbar^2}{2m} \frac{\partial^2}{\partial x^2} + V(x) \right] \psi(x,t) = i\hbar \frac{\partial}{\partial t} \psi(x,t). \tag{3.3}$$

The wave function $\psi(x,t)$ is a function of position x and time t. This equation is a *partial* differential equation in the independent variables x and t. To solve it we employ the method of separation of variables. This technique makes use of the fact that the operator on the left-hand side of Eq. (3.3) depends on x only, whereas the operator on the right-hand side depends on t only. This implies a product solution of the form

$$\psi(x,t) = u(x)f(t).$$

With this substitution for $\psi(x,t)$ Eq. (3.3) becomes

$$\frac{1}{u(x)} \left[\frac{-\hbar^2}{2m} \frac{d^2}{dx^2} u(x) + V(x)u(x) \right] = \frac{i\hbar}{f(t)} \frac{d}{dt} f(t). \tag{3.4}$$

Partial derivatives have been replaced by total derivatives, since $u(x)$ and $f(t)$ are each functions of single variables. We see that the left-hand side of Eq. (3.4) depends only on x, whereas the right-hand side depends only on t. However, x and t vary independently and the equation must hold for all x and t. For example, if we hold x constant, then the left-hand side remains constant regardless of how we vary t. Thus, both sides of Eq. (3.4) must be equal to the same constant for all x and t. Denote the constant by E. Then,

$$\frac{i\hbar}{f(t)} \frac{d}{dt} f(t) = E,$$

which we can integrate directly to obtain the solution

$$f(t) = f_0 e^{-iEt/\hbar}.$$

The constant E represents the total energy of the system.

After some rearrangement we also get from Eq. (3.4)

$$\frac{d^2}{dx^2} u(x) + \frac{2m}{\hbar^2} [E - V(x)] u(x) = 0. \tag{3.5}$$

Therefore, the solutions to Eq. (3.3) are of the form

$$\psi(x, t) = u(x) e^{-iEt/\hbar},$$

where the function $u(x)$ satisfies the second-order differential equation (Eq. 3.5).

For a harmonic oscillator with potential energy

$$V(x) = \tfrac{1}{2} m\omega^2 x^2,$$

Eq. (3.5) becomes

$$\frac{d^2}{dx^2} u(x) + \left[\frac{2mE}{\hbar^2} - \frac{m^2\omega^2}{\hbar^2} x^2 \right] u(x) = 0.$$

By a change of dependent variable this equation can be transformed into the confluent hypergeometric equation.

It is convenient to introduce the dimensionless variable $\rho = x/b$. The constant parameter b has the dimension of a length. This gives

$$\frac{d^2}{d\rho^2} u(\rho) + (\mu - \rho^2) u(\rho) = 0. \tag{3.6}$$

Here we have chosen $b = (\hbar/m\omega)^{1/2}$ and defined $\mu = 2E/\hbar\omega$.

Next, we look for a change of dependent variable that will effect the transformation of Eq. (3.6) to the form of Eq. (3.1) or possibly Eq. (2.24). In either case the zero-order term contains only the dependent variable multiplied by a constant. With this in mind we try the substitution

$$u(\rho) = f(\rho) e^{g(\rho)} \tag{3.7}$$

where $f(\rho)$ and $g(\rho)$ are functions of ρ to be determined. This change is motivated by the fact that successive differentiations always give an overall factor e^g, which can be cancelled out of the differential equation leaving a zero-order term consisting of $f(\rho)$ multiplied by an arbitrary function of ρ. We can choose this latter function to be equal to a constant to get the right form for the zero-order term.

By substituting Eq. (3.7) into Eq. (3.6), multiplying through by e^{-g}, and collecting terms, we obtain a differential equation for $f(\rho)$,

$$f'' + 2g'f' + (\mu - \rho^2 + g'^2 + g'')f = 0.$$

In the zero-order term the factor in parentheses should be a constant. This is easily accomplished, if we take $g'^2 = \rho^2$. In this case $g' = \pm\rho$, $g = \pm\frac{1}{2}\rho^2$, and $g'' = \pm 1$. We choose the sign so that the solution to Eq. (3.6) does not diverge for large values of ρ. Equation (3.7) then reads

$$u(\rho) = f(\rho)e^{-\frac{1}{2}\rho^2}.$$

The function $f(\rho)$ satisfies the differential equation,

$$f''(\rho) - 2\rho f'(\rho) + (\mu - 1)f(\rho) = 0. \tag{3.8}$$

Although the zero-order term in Eq. (3.8) is in the form we want, the other two terms are not. We note, however, that a change in *independent* variable can alter the first two terms without affecting the zero-order term. Thus, we look for a change in independent variable that will yield the desired form for Eq. (3.8). We try the change

$$s = \alpha\rho^n$$

where α and n are parameters to be determined. The differential operators transform as

$$\frac{d}{d\rho} = n\alpha^{1/n}s^{1-1/n}\frac{d}{ds}$$

and

$$\frac{d^2}{d\rho^2} = n^2\alpha^{2/n}s^{2-2/n}\frac{d^2}{ds^2} + n(n-1)\alpha^{2/n}s^{1-2/n}\frac{d}{ds}.$$

By inspection we see that $n = 2$ gives a factor s in the second-order term as is required to fit the form of Eq. (3.1). Then by choosing $\alpha = 1$ Eq. (3.8) becomes

$$sf''(s) + \left(\frac{1}{2} - s\right)f'(s) - \frac{1}{4}(1 - \mu)f(s) = 0,$$

with $s = \rho^2$. This is the confluent hypergeometric equation. According to Eq. (3.2), the general solution is

$$f(\rho) = A\,_1F_1\left(\frac{1}{4}[1 - \mu]; \frac{1}{2}; \rho^2\right) + B\rho\,_1F_1\left(\frac{1}{4}[3 - \mu]; \frac{3}{2}; \rho^2\right). \tag{3.9}$$

3.2.1 BOUNDARY CONDITIONS AND ENERGY EIGENVALUES

We still have the physical constraint that when the particle is far from the center of force, the solution $u(\rho)$ must not diverge. That is, $u(\rho)$ must remain finite as $|\rho|$ tends to infinity. In view of this we examine the asymptotic behavior of $f(\rho)$. Let us look separately at the two confluent hypergeometric functions in Eq. (3.9). For the first one, write

$$
{}_1F_1\left(\tfrac{1}{4}[1-\mu];\tfrac{1}{2};\rho^2\right) = \frac{\Gamma\left(\tfrac{1}{2}\right)}{\Gamma\left(\tfrac{1}{4}(1-\mu)\right)} \sum_{k=0}^{\infty} \frac{\Gamma\left(k+\tfrac{1}{4}[1-\mu]\right)}{k!\,\Gamma\left(k+\tfrac{1}{2}\right)}\rho^{2k}.
$$

For large k (specifically, $k >> \tfrac{1}{4}(1-\mu)$ and $k >> \tfrac{1}{2}$) both $\Gamma\left(k+\tfrac{1}{4}[1-\mu]\right)$ and $\Gamma\left(k+\tfrac{1}{2}\right)$ are approximated by $\Gamma(k)$ so that the terms for large k in the sum are approximated by $\rho^{2k}/k!$.

Thus, by adding and then subtracting terms for small k, we have the approximation[3]

$$
{}_1F_1\left(\tfrac{1}{4}[1-\mu];\tfrac{1}{2};\rho^2\right) \approx \frac{\Gamma\left(\tfrac{1}{2}\right)}{\Gamma\left(\tfrac{1}{4}(1-\mu)\right)} \left(\sum_{k=0}^{\infty} \frac{\rho^{2k}}{k!} \right.
$$
$$
\left. -[\text{polynomial in } \rho^2] \right).
$$

The first term on the right-hand side is the series expansion for e^{ρ^2}, which dominates for large values of $|\rho|$. Thus, we have the behavior

$$
{}_1F_1\left(\tfrac{1}{4}[1-\mu];\tfrac{1}{2};\rho^2\right) \sim e^{\rho^2} \quad \text{for } |\rho| >> 1. \tag{3.10a}
$$

By a similar argument we find that

$$
{}_1F_1\left(\tfrac{1}{4}[3-\mu];\tfrac{3}{2};\rho^2\right) \sim e^{\rho^2} \quad \text{for } |\rho| >> 1. \tag{3.10b}
$$

With these results the asymptotic behavior of $u(\rho)$ in Eq. (3.7) is

$$
u(\rho) = e^{-\frac{1}{2}\rho^2} f(\rho) \xrightarrow[|\rho|\to\infty]{} (A' + \rho B')e^{\frac{1}{2}\rho^2}.
$$

To avoid this divergence for large $|\rho|$ we must require that the series terminate. This means that in Eq. (3.9) either

a. $-\tfrac{1}{4}(1-\mu)$ is a nonnegative integer and $B = 0$ or

b. $-\tfrac{1}{4}(3-\mu)$ is a nonnegative integer and $A = 0$.

---

[3] An improved and more general approximation is derived in Section 10.8.

The first case gives for the energy of the particle

$$E_n = \left(n + \tfrac{1}{2}\right)\hbar\omega, \qquad n = 0, 2, 4, \ldots$$

with

$$u_n(\rho) = A_n \, {}_1F_1\left(-\tfrac{n}{2}; \tfrac{1}{2}; \rho^2\right)e^{-\frac{1}{2}\rho^2}.$$

In obtaining the energy we have made use of the definition of μ. The second case gives

$$E_n = \left(n + \tfrac{1}{2}\right)\hbar\omega, \qquad n = 1, 3, 5, \ldots$$

with

$$u_n(\rho) = B_n\rho \, {}_1F_1\left(-\tfrac{1}{2}[n-1]; \tfrac{3}{2}; \rho^2\right)e^{-\frac{1}{2}\rho^2}.$$

Note that if n is even, then ${}_1F_1\left(-\tfrac{n}{2}; \tfrac{1}{2}; \rho^2\right)$ is a polynomial in even powers of ρ. For odd n, the function $\rho\, {}_1F_1\left(-\tfrac{1}{2}[n-1]; \tfrac{3}{2}; \rho^2\right)$ is a polynomial in odd powers of ρ.

3.2.2   HERMITE POLYNOMIALS AND THE CONFLUENT HYPERGEOMETRIC FUNCTION

With a change of summation index in the confluent hypergeometric series we can write for even n

$${}_1F_1\left(-\tfrac{n}{2}; \tfrac{1}{2}; \rho^2\right) = \sum_{k=0}^{n/2} \frac{\left(-\tfrac{1}{2}n\right)_{\frac{1}{2}n-k}}{\left(\tfrac{1}{2}n - k\right)!\left(\tfrac{1}{2}\right)_{\frac{1}{2}n-k}} \rho^{n-2k}.$$

If we use the identities

$$(n - 2k)! = 2^{n-2k}\left(\tfrac{1}{2}\right)_{\frac{1}{2}n-k}\left(\tfrac{1}{2}n - k\right)!$$

and

$$\left(-\tfrac{n}{2}\right)_{\frac{1}{2}n-k} = \frac{(-1)^{\frac{1}{2}n-k}\left(\tfrac{n}{2}\right)!}{k!},$$

it follows that

$${}_1F_1\left(-\tfrac{n}{2}; \tfrac{1}{2}; \rho^2\right) = \frac{(-1)^{\frac{1}{2}n}\left(\tfrac{1}{2}n\right)!}{n!} \sum_{k=0}^{n/2} \frac{(-1)^k n!}{k!(n - 2k)!}(2\rho)^{n-2k}. \qquad (3.11a)$$

In a similar fashion for odd n we also find that

$$\rho\, {}_1F_1\left(-\tfrac{n-1}{2}; \tfrac{3}{2}; \rho^2\right) = \frac{(-1)^{\frac{1}{2}(n-1)}\left(\tfrac{n-1}{2}\right)!}{2n!} \sum_{k=0}^{\frac{n-1}{2}} \frac{(-1)^k n!}{k!(n - 2k)!}(2\rho)^{n-2k}.$$

$$(3.11b)$$

We define the functions $H_n(\rho)$ by

$$H_n(\rho) = \frac{n!(-1)^{-\frac{1}{2}n}}{\left(\tfrac{n}{2}\right)!}\, {}_1F_1\left(-\tfrac{n}{2}; \tfrac{1}{2}; \rho^2\right) \quad \text{for even } n, \qquad (3.12)$$

and

$$H_n(\rho) = \frac{2n!(-1)^{\frac{1}{2}(1-n)}}{\left(\frac{n-1}{2}\right)!} \rho\, {}_1F_1\left(-\tfrac{n-1}{2}; \tfrac{3}{2}; \rho^2\right) \quad \text{for odd } n. \tag{3.13}$$

These functions are called *Hermite polynomials*. From Eq. (3.8) we see that they are solutions of the differential equation

$$H_n''(\rho) - 2\rho H_n'(\rho) + 2n H_n(\rho) = 0, \tag{3.14}$$

which is known as *Hermite's* equation. In subsequent chapters we examine the properties of these functions in some detail. With the normalization chosen we see from Eqs. (3.11) that we can write[4] for *any* integer n

$$H_n(\rho) = \sum_{k=0}^{[n/2]} \frac{(-1)^k n!}{k!(n-2k)!} (2\rho)^{n-2k}. \tag{3.15}$$

On putting the pieces together the harmonic oscillator eigenfunctions take the form[5]

$$u_n(\rho) = N_n H_n(\rho) e^{-\frac{1}{2}\rho^2}, \tag{3.16}$$

where N_n is a normalization constant to be determined later (see Eq. 12.29).

EXERCISES

1. Show that

$$b(b+1)_k = (b+k)(b)_k$$
$$= (b)_{k+1},$$

and use these results to prove the following:

a. $\dfrac{d}{dz}\, {}_1F_1(a; c; z) = \dfrac{a}{c}\, {}_1F_1(a+1; c+1; z),$

b. $z\dfrac{d}{dz}\, {}_1F_1(a; c; z) = a\, {}_1F_1(a+1; c; z) - a\, {}_1F_1(a; c; z),$

c. ${}_1F_1(a; c; z) = {}_1F_1(a-1; c; z) + \dfrac{z}{c}\, {}_1F_1(a; c+1; z)$

---

[4]We use the bracket notation $[x]$ to denote the largest integer in x.

[5]Merzbacher presents a nice treatment of the *double* oscillator in terms of confluent hypergeometric functions, which he uses to give a qualitative description of the behavior of the nitrogen atom in the ammonia molecule. E. Merzbacher, *Quantum Mechanics*, J. Wiley and Sons, New York, 1961, p. 194.

2. For the differential equation

$$z^2 u''(z) - z^2 u'(z) + \frac{3}{16} u(z) = 0$$

try solutions of the form

$$u(z) = z^s f(z)$$

where s is a number. Find the two values of s such that the differential equation for $f(z)$ is the confluent hypergeometric equation. For each of these values of s write down the confluent hypergeometric equation and the corresponding confluent hypergeometric function. Write down explicitly the general solution to the equation for $u(x)$.

3. For the differential equation

$$xu''(x) + (1 - 5x)u'(x) + 6xu(x) = 0,$$

try the solution,

$$u(x) = e^{\mu x} f(x).$$

Choose a value for μ such that the differential equation for $f(x)$ is the confluent hypergeometric equation. Write down the solution $u(x)$ for which $f(x)$ is a polynomial.

4. Transform the differential equation

$$f''(x) - f'(x) - \frac{\mu(\mu - 1)}{x^2} f(x) = 0$$

by means of the change of dependent variable,

$$f(x) = u(x)v(x).$$

Find the function $v(x)$ such that $u(x)$ is the solution to the confluent hypergeometric equation. Express the general solution $f(x)$ in terms of confluent hypergeometric functions.

5. Solutions to the linear second-order differential equation

$$4zu''(z) + 2u'(z) - zu(z) = 0$$

may be written as

$$u(z) = e^{\gamma z} f(z).$$

a. Find a value for γ such that the differential equation for $f(z)$ is the confluent hypergeometric equation. From these results write down explicitly the general solution to the equation for $u(z)$.

b. Solve the differential equation for $u(z)$ by assuming a series solution,

$$u(z) = \sum_{k=0}^{\infty} c_k z^{k+s}.$$

Find the explicit expression for the general coefficient c_k. Can both c_0 and c_1 be zero at the same time? Explain. What values may the parameter s have? Under what conditions? Choose one of the values for s and show that the corresponding solution can be written in terms of the generalized hypergeometric function,

$$u(z) = z^{\mu} \,{}_0F_1\left(\nu; \left(\frac{z}{4}\right)^2\right),$$

where μ and ν are numbers which you should determine explicitly.

6. Solve the differential equation

$$u''(x) - u'(x) - \frac{1}{4x^2}[\mu(\mu-2) + 2\mu x]u(x) = 0$$

by assuming that the solution can be expressed as

$$u(x) = v(x)w(x).$$

Choose the function $v(x)$ so that $w(x)$ satisfies the confluent hypergeometric equation. Write the general solution $u(x)$ in terms of confluent hypergeometric functions. Show that a particular solution reduces to $x^{\frac{1}{2}\mu}e^x$.

7. Assume that the differential equation

$$z^2 f''(z) - z^2 f'(z) + \frac{1}{2}\left(z - \frac{3}{2}\right) f(z) = 0$$

has a solution which can be written as

$$f(z) = u(z)v(z).$$

Find $v(z)$ such that $u(z)$ is a solution to the confluent hypergeometric equation. Show that one of the solutions reduces to

$$f(z) = z^{-1/2}(e^z - z - 1).$$

8. Show by iteration of Eq. (3.15) that

$$x^n = \frac{1}{2^n} \sum_{k=0}^{[n/2]} \frac{n!}{k!(n-k)!} H_{n-2k}(x).$$

9. From Leibniz's theorem we find that[6]

$$\frac{d^m}{dx^m}[x^n e^x] = e^x \sum_{k=0}^{n} \frac{(m-k+1)_k (n-k+1)_k}{k!} x^{n-k}$$

where m and n are positive integers with $m > n$. Show that

$$\frac{d^m}{dx^m}[x^n e^x] = \frac{m!}{(m-n)!} e^x {}_1F_1(-n; m-n+1; -x).$$

---

[6] See Exercise 2.27.

4

Problems in Two Dimensions

4.1 Cylindrical Wave Guides

Transmission of electromagnetic energy may be accomplished by means of hollow metallic cylinders called wave guides. The electromagnetic waves propagate in the nonconducting region enclosed by the metal cylinder. For the fields in this region, Maxwell's equations reduce to[1]

$$\nabla \cdot \mathbf{E} = 0, \tag{4.1a}$$

$$\nabla \times \mathbf{E} + \mu \frac{\partial \mathbf{H}}{\partial t} = 0, \tag{4.1b}$$

$$\nabla \cdot \mathbf{H} = 0, \tag{4.1c}$$

$$\nabla \times \mathbf{H} - \varepsilon \frac{\partial \mathbf{E}}{\partial t} = 0. \tag{4.1d}$$

The vectors $\mathbf{E}$ and $\mathbf{H}$ are the electric and magnetic fields, respectively. We have assumed that the nonconducting medium is linear, homogeneous, isotropic, and nonabsorbing, and contains no free electric charge.

For a guide with a circular cross section the geometry is shown in Figure 4.1 where the position vector $\mathbf{r}$ is represented in terms of rectangular coordinates (x, y, z) and cylindrical coordinates (ρ, θ, z).

If we take the curl of Eq. (4.1b) and use of the identity

$$\nabla \times \nabla \times \mathbf{F} = -\nabla^2 \mathbf{F} + \nabla(\nabla \cdot \mathbf{F}),$$

where $\mathbf{F}$ is any vector, we obtain from Maxwell's equations

$$\nabla^2 \mathbf{E} - \mu\varepsilon \frac{\partial^2 \mathbf{E}}{\partial t^2} = 0. \tag{4.2a}$$

A similar operation on Eq. (4.1d) leads to

$$\nabla^2 \mathbf{H} - \mu\varepsilon \frac{\partial^2 \mathbf{H}}{\partial t^2} = 0. \tag{4.2b}$$

---

[1]In this book we use SI units. See, for example, P. Lorrain and D. Corson, *Electromagnetic Fields and Waves*, W.H. Freeman and Company, San Francisco, 1970.

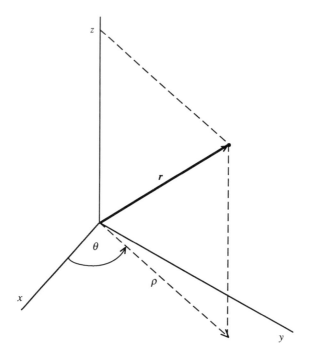

Figure 4.1

Since the space and time operators (∇^2 and $\partial^2/\partial t^2$, respectively) are completely separable, each solution to Eq. (4.2a) (or Eq. 4.2b) is a product of a space dependent part and a time dependent part.[2]

For monochromatic waves (of frequency $\nu = \omega/2\pi$ and wavelength $\lambda = 2\pi/k$) propagating in the z-direction it is easily verified by substitution into the equations that the solutions to Eqs. (4.2a) and (4.2b) are, respectively[3]

$$\mathbf{E}(\mathbf{r},t) = \mathbf{E}(\rho,\theta)e^{i(kz-\omega t)}, \tag{4.3a}$$

$$\mathbf{H}(\mathbf{r},t) = \mathbf{H}(\rho,\theta)e^{i(kz-\omega t)}. \tag{4.3b}$$

The transverse components of $\mathbf{E}$ and $\mathbf{H}$ can be expressed in terms of derivatives of the longitudinal components E_z and H_z.[4] Therefore, all components of $\mathbf{E}$ and $\mathbf{H}$ can be constructed from solutions of the differential equations for E_z and H_z. It is sufficient to consider these equations only.

---

[2]See Section 3.2 for the solution of partial differential equations by the method of separation of variables.

[3]Here and throughout the chapter, i denotes the imaginary number $\sqrt{-1}$. For a discussion of complex numbers see Chapter 7.

[4]For example, see P. Lorrain and D. Corson, *op. cit.*, Chapter 13.

The solutions to Maxwell's equations must satisfy certain boundary conditions at the wall of the wave guide. We distinguish two types of fields: *transverse magnetic* (TM) and *transverse electric* (TE). If the wall is a perfect conductor, then the boundary conditions for TM waves are

TM waves: $H_z(\mathbf{r}) = 0$ *everywhere* in the wave guide,

$E_z(\mathbf{r}) = 0$ *on inside wall* of the wave guide.

In cylindrical coordinates the differential operator ∇^2 is

$$\nabla^2 = \frac{1}{\rho}\frac{\partial}{\partial\rho}\left(\rho\frac{\partial}{\partial\rho}\right) + \frac{1}{\rho^2}\frac{\partial^2}{\partial\theta^2} + \frac{\partial^2}{\partial z^2}. \tag{4.4}$$

On using this operator in Eq. (4.2a) with solutions of the type given in Eq. (4.3a) we find that TM waves in a cylindrical wave guide are described by the differential equation

$$\left[\frac{\partial^2}{\partial\rho^2} + \frac{1}{\rho}\frac{\partial}{\partial\rho} + \frac{1}{\rho^2}\frac{\partial^2}{\partial\theta^2} + \gamma^2\right]E_z(\rho,\theta) = 0. \tag{4.5}$$

The constant γ^2 is defined by

$$\gamma^2 = \mu\varepsilon\omega^2 - k^2. \tag{4.6}$$

On multiplying Eq. (4.5) by ρ^2 we get the equation,

$$\left[\rho^2\frac{\partial^2}{\partial\rho^2} + \rho\frac{\partial}{\partial\rho} + \gamma^2\rho^2 + \frac{\partial^2}{\partial\theta^2}\right]E_z(\rho,\theta) = 0, \tag{4.7}$$

in which the operator is separated into a ρ-dependent part and a θ-dependent part. This allows us to use the method of separation of variables to solve the partial differential equation for $E_z(\rho,\theta)$. We look for solutions that are products of the type

$$E_z(\rho,\theta) = u(\rho)v(\theta). \tag{4.8}$$

If we make this substitution for $E_z(\rho,\theta)$ in Eq. (4.7), we find that

$$\frac{1}{u}\left[\rho^2\frac{d^2u}{d\rho^2} + \rho\frac{du}{d\rho} + \gamma^2\rho^2u\right] = -\frac{1}{v}\frac{d^2v}{d\theta^2}. \tag{4.9}$$

Since the left-hand side depends on ρ only, the right-hand side on θ only, and ρ and θ are independent, each side must be equal to a constant, say β^2. The right-hand side of Eq. (4.9) then leads to

$$\frac{d^2v(\theta)}{d\theta^2} + \beta^2v(\theta) = 0,$$

which has solutions

$$v(\theta) = Ae^{i\beta\theta} + Be^{-i\beta\theta}.$$

Since the field has a definite value at each point in space, $E_z(\rho, \theta)$ must be a single-valued function of θ. An increase in θ by some integral multiple of 2π brings us back to the same spatial point. This means that we must have

$$v(\theta + 2\pi) = v(\theta),$$

which requires that

$$A = Ae^{i2\pi\beta} \quad \text{and} \quad B = Be^{-i2\pi\beta}.$$

This holds only if β is an integer.

From the left-hand side of Eq. (4.9), we get

$$\rho^2 \frac{d^2u}{d\rho^2} + \rho \frac{du}{d\rho} + (\gamma^2\rho^2 - n^2)u = 0, \tag{4.10}$$

where n is an integer.

To express this equation in dimensionless form we define a new independent variable $x = \gamma\rho$, which transforms Eq. (4.10) into

$$x^2 \frac{d^2u}{dx^2} + x \frac{du}{dx} + (x^2 - n^2)u = 0. \tag{4.11}$$

This is *Bessel's* equation. It has two singularities,[5] a regular one at $x = 0$ and an irregular one at $x = \infty$, which suggests that we try by a change of variable to cast it into the form of the confluent hypergeometric equation.

4.1.1 BESSEL'S EQUATION AND THE CONFLUENT HYPERGEOMETRIC EQUATION

Based on our experience with Hermite's equation in the preceding chapter, we change the dependent variable according to

$$u(x) = x^s e^{g(x)} f(x), \tag{4.12}$$

where the number s and the functions $f(x)$ and $g(x)$ are to be determined. We make this substitution in Eq. (4.11) and collect terms to get

$$x^s e^g \big[x^2 f''(x) + (2sx + 2g'x^2 + x)f'(x)$$

$$+ (g'^2x^2 + 2sg'x + s(s-1) + g''x^2 + xg' + s + x^2 - n^2)f(x) \big] = 0. \tag{4.13}$$

A comparison of this equation with Eq. (3.1) shows that we need to have

$$g'^2x^2 + 2g'sx + s^2 + xg' + x^2 - n^2 + g''x^2 = -ax$$

---

[5]See Exercise 4.3.

where a is a constant. Clearly, this can be achieved if

$$s^2 - n^2 = 0,$$
$$g'' + g'^2 + 1 = 0,$$
$$(2s + 1)g' = -a.$$

From the first of these equations we see that $s = \pm n$. The second implies that $g' = \pm i$. To get the standard form of the confluent hypergeometric equation we take $s = n$ and $g' = -i$. With this choice Eq. (4.13) can be written as

$$x f''(x) + (2n + 1 - 2ix) f'(x) - i(2n + 1) f(x) = 0. \qquad (4.14)$$

Our task is almost complete. A further change in independent variable of the form $z = \alpha x$ gives

$$z \frac{d^2 f}{dz^2} + \left(2n + 1 - \frac{2iz}{\alpha} \right) \frac{df}{dz} - \frac{2i}{\alpha} \left(n + \tfrac{1}{2} \right) f = 0.$$

Finally, we choose the constant $\alpha = 2i$ to obtain

$$z f''(z) + (2n + 1 - z) f'(z) - \left(n + \tfrac{1}{2} \right) f(z) = 0, \qquad (4.15)$$

which is exactly of the form of the confluent hypergeometric equation. Thus,

$$f(x) = A \, {}_1F_1 \left(n + \tfrac{1}{2}; 2n + 1; 2ix \right) + B(2ix)^{-2n} \, {}_1F_1 \left(-n + \tfrac{1}{2}; -2n + 1; 2ix \right),$$

and the solution to Eq. (4.11) is

$$u(x) = e^{-ix} \left[A_n x^n \, {}_1F_1 \left(n + \tfrac{1}{2}; 2n + 1; 2ix \right) \right.$$

$$\left. + B_n x^{-n} \, {}_1F_1 \left(-n + \tfrac{1}{2}; -2n + 1; 2ix \right) \right], \qquad (4.16)$$

where A_n and B_n are arbitrary constants.

Note that in the transformation from Eq. (4.11) to Eq. (4.15) we have not made use of the fact that n is an integer. Indeed, in general, n need not be an integer, although in the application we are making here to wave guides it must be.

4.1.2   BESSEL FUNCTIONS OF ARBITRARY ORDER

For the moment let us digress a bit and consider the more general case with $n = \nu$ where ν is *not* necessarily an integer. In the general solution given in Eq. (4.16) we define new constants of integration by

$$a_\nu = 2^\nu \, \Gamma(\nu + 1) A_\nu \quad \text{and} \quad b_\nu = 2^{-\nu} \Gamma(-\nu + 1) B_\nu.$$

The first term in Eq. (4.16) now becomes

$$a_\nu \frac{e^{-ix}}{\Gamma(\nu+1)} \left(\frac{x}{2}\right)^\nu {}_1F_1\left(\nu+\tfrac{1}{2};2\nu+1;2ix\right) = a_\nu J_\nu(x)$$

with the definition,

$$J_\nu(x) = \frac{e^{-ix}}{\Gamma(\nu+1)} \left(\frac{x}{2}\right)^\nu {}_1F_1\left(\nu+\tfrac{1}{2};2\nu+1;2ix\right). \qquad (4.17)$$

For the second term in Eq. (4.16) we have

$$b_\nu \frac{e^{-ix}}{\Gamma(-\nu+1)} \left(\frac{x}{2}\right)^{-\nu} {}_1F_1\left(-\nu+\tfrac{1}{2};-2\nu+1;2ix\right) = b_\nu J_{-\nu}(x).$$

Thus, the general solution of Eq. (4.11) is written more compactly as

$$u(x) = a_\nu J_\nu(x) + b_\nu J_{-\nu}(x). \qquad (4.18)$$

The function $J_\nu(x)$ defined by Eq. (4.17) is called a *Bessel function of the first kind* of order ν.

4.1.3   FORMULAS INVOLVING BESSEL FUNCTIONS

A frequently encountered series expansion of $J_\nu(x)$ is

$$J_\nu(x) = \left(\frac{x}{2}\right)^\nu \sum_{m=0}^{\infty} \frac{(-1)^m}{m!\,\Gamma(m+\nu+1)} \left(\frac{x}{2}\right)^{2m}$$

$$= \frac{1}{\Gamma(\nu+1)} \left(\frac{x}{2}\right)^\nu {}_0F_1(\nu+1;-x^2/4), \qquad (4.19)$$

which converges for all x. This series is used by some writers to define the Bessel function.[6]

With this representation we can evaluate the integral

$$I = \int_0^\infty e^{-r^2 y^2} J_\nu(sy) y^{\nu+1} dy,$$

where r and s are positive numbers and $\nu > -1$. On substituting the expansion in Eq. (4.19) for the Bessel function and interchanging the order of summation and integration,[7] we get

$$I = \sum_{k=0}^{\infty} \frac{(-1)^k}{k!\,\Gamma(k+\nu+1)} \left(\frac{s}{2}\right)^{2k+\nu} \int_0^\infty y^{2k+2\nu+1} e^{-r^2 y^2} dy.$$

---

[6]In Chapter 9 as an exercise in contour integration, we show the equivalence of the two representations of $J_\nu(x)$ given by Eqs. (4.17) and (4.19).

[7]We assume that in cases of physical interest such interchanges are always permissible.

The integral here can be evaluated using the formula[8]

$$\int_0^\infty e^{-a^2 t^2} t^b \, dt = \frac{\Gamma\left(\frac{1}{2}(b+1)\right)}{2a^{b+1}}.$$

Finally, we have for the integral I

$$\int_0^\infty e^{-r^2 y^2} J_\nu(sy) y^{\nu+1} \, dy = \frac{s^\nu}{(2r^2)^{\nu+1}} e^{-s^2/4r^2}. \tag{4.20}$$

Some useful recursion formulas are readily obtained from the series expansion of $J_\nu(x)$. From Eq. (4.19) we have

$$J_{\nu-1}(x) = \sum_{k=0}^\infty \frac{(-1)^k (k+\nu)}{k! \Gamma(k+\nu+1)} \left(\frac{x}{2}\right)^{2k+\nu-1}$$

$$= -\sum_{k=1}^\infty \frac{(-1)^{k-1}}{(k-1)! \Gamma(k+\nu+1)} \left(\frac{x}{2}\right)^{2(k-1)+\nu+1}$$

$$+ \nu \sum_{k=0}^\infty \frac{(-1)^k}{k! \Gamma(k+\nu+1)} \left(\frac{x}{2}\right)^{2k+\nu-1}$$

$$= -J_{\nu+1}(x) + \frac{2\nu}{x} J_\nu(x).$$

Thus,

$$J_{\nu-1}(x) + J_{\nu+1}(x) = \frac{2\nu}{x} J_\nu(x). \tag{4.21}$$

If we differentiate the series in Eq. (4.19) term by term, we see that

$$\frac{d}{dx} J_\nu(x) = \frac{1}{2} \sum_{k=0}^\infty \frac{(-1)^k (2k+\nu)}{k! \Gamma(k+\nu+1)} \left(\frac{x}{2}\right)^{2k+\nu-1}$$

$$= -\sum_{k=1}^\infty \frac{(-1)^{k-1}}{(k-1)! \Gamma(k+\nu+1)} \left(\frac{x}{2}\right)^{2(k-1)+\nu+1}$$

$$+ \frac{\nu}{2} \sum_{k=0}^\infty \frac{(-1)^k}{k! \Gamma(k+\nu+1)} \left(\frac{x}{2}\right)^{2k+\nu-1}$$

$$= -J_{\nu+1}(x) + \frac{\nu}{x} J_\nu(x),$$

or

$$J_\nu'(x) = \frac{\nu}{x} J_\nu(x) - J_{\nu+1}(x). \tag{4.22}$$

By combining Eqs. (4.21) and (4.22) we obtain

$$J_\nu'(x) = \frac{-\nu}{x} J_\nu(x) + J_{\nu-1}(x). \tag{4.23}$$

---

[8] See Exercise 1.14.

The addition of Eqs. (4.22) and (4.23) yields

$$J_{\nu-1}(x) - J_{\nu+1}(x) = 2J'_\nu(x). \tag{4.24}$$

Let us consider the case where $\nu > 0$, but ν is *not* an integer. We see from Eq. (4.19) that for $|x| \ll 1$,

$$J_\nu(x) = \left(\frac{x}{2}\right)^\nu \left[\frac{1}{\Gamma(\nu+1)} + O(x^2)\right] \xrightarrow[x\to 0]{} 0, \tag{4.25a}$$

$$J_{-\nu}(x) = \left(\frac{x}{2}\right)^{-\nu} \left[\frac{1}{\Gamma(-\nu+1)} + O(x^2)\right] \xrightarrow[x\to 0]{} \infty. \tag{4.25b}$$

Clearly, $J_\nu(x)$ is a solution of Eq. (4.11) that is regular at $x = 0$, whereas the solution $J_{-\nu}(x)$ is irregular at that point.

Now let us examine the function $J_{-\nu}(x)$ for the case with $\nu = n$ where n is an integer. We use the representation in Eq. (4.19) to write

$$J_{-n}(x) = \sum_{k=0}^\infty \frac{(-1)^k}{k!\Gamma(k-n+1)} \left(\frac{x}{2}\right)^{2k-n}.$$

In the first n terms of this sum, k is an integer which is less than n. Now, the gamma function is singular if its argument is a negative integer or zero.[9] That is,

$$\Gamma(k-n+1) = \infty \quad \text{for} \quad k \le n-1.$$

Thus, the first n terms of the sum vanish, and we have

$$J_{-n}(x) = \sum_{k=n}^\infty \frac{(-1)^k}{k!\Gamma(k-n+1)} \left(\frac{x}{2}\right)^{2k-n}.$$

We define a new summation index by $m = k - n$ and rewrite this sum as

$$J_{-n}(x) = (-1)^n \left(\frac{x}{2}\right)^n \sum_{m=0}^\infty \frac{(-1)^m}{m!\Gamma(m+n+1)} \left(\frac{x}{2}\right)^{2m},$$

$$= (-1)^n J_n(x).$$

In obtaining this last result we have twice used the relation $\Gamma(k+1) = k!$, which holds for any nonnegative integer k. We see that if ν is an integer, then the two functions $J_\nu(x)$ and $J_{-\nu}(x)$ are *not* linearly independent, in which case Eq. (4.18) is not a *general* solution of Eq. (4.11).

---

[9]Refer to Exercise 1.10 or to Eq. (10.5).

4.1.4 LINEARLY INDEPENDENT SOLUTIONS OF BESSEL'S EQUATION

A function that *is* linearly independent with respect to $J_\nu(x)$ and also satisfies Eq. (4.11) is defined by[10]

$$N_\nu(x) = \frac{J_\nu(x) \cos \nu\pi - J_{-\nu}(x)}{\sin \nu\pi}. \tag{4.26}$$

This solution to Eq. (4.11) is called a *Bessel function of the second kind*.[11] It has the virtue that, even in the limit $\nu \to$ *integer*, it remains a valid solution of Eq. (4.11) and does not depend linearly on $J_\nu(x)$. Whether ν is an integer or not, the function $N_\nu(x)$ is irregular at $x = 0$.

For all $\nu \geq 0$ the solution $J_\nu(x)$ is regular at the origin, whereas the solution $N_\nu(x)$ is irregular at that point.

Two linear combinations of $J_\nu(x)$ and $N_\nu(x)$ frequently encountered in physical applications[12] are

$$H_\nu^{(1)}(x) = J_\nu(x) + iN_\nu(x) \tag{4.27}$$

and

$$H_\nu^{(2)}(x) = J_\nu(x) - iN_\nu(x). \tag{4.28}$$

These functions, known as *Hankel functions*[13] of the first and second kind, respectively, are linearly independent solutions of Bessel's equation.

4.1.5 AN INTEGRAL REPRESENTATION OF $J_\nu(x)$

An integral representation of $J_\nu(x)$ may be obtained directly from the series expansion in Eq. (4.19). First, we use the identity[14]

$$(2m)! = \frac{1}{\sqrt{\pi}} 2^{2m} \Gamma\left(m + \tfrac{1}{2}\right) m!$$

to substitute for $m!$ in Eq. (4.19). Then, we multiply both sides of Eq. (4.19) by $\Gamma\left(\nu + \tfrac{1}{2}\right)$ to get

$$\Gamma\left(\nu + \tfrac{1}{2}\right) J_\nu(x) = \frac{1}{\sqrt{\pi}} \left(\frac{x}{2}\right)^\nu \sum_{m=0}^\infty \frac{(-1)^m}{(2m)!} \frac{\Gamma\left(\nu + \tfrac{1}{2}\right)\Gamma\left(m + \tfrac{1}{2}\right)}{\Gamma(\nu + m + 1)} x^{2m}.$$

---

[10] A rationale for this definition may be seen by comparing the asymptotic forms of $J_\nu(x)$ and $N_\nu(x)$ for large values of x. See Exercises 10.8 and 10.9. For a more complete discussion of the history of these functions, see G.N. Watson, *op. cit.*, p. 63 *ff.*

[11] Another name that is often used for this function is the *Neumann function*.

[12] An example occurs in Exercise 10.12.

[13] These functions are sometimes referred to as *Bessel functions of the third kind*.

[14] See Exercise 1.7.

Now using Eq. (1.24) we obtain

$$\Gamma\left(\nu + \tfrac{1}{2}\right) J_\nu(x) = \frac{2}{\sqrt{\pi}}\left(\frac{x}{2}\right)^\nu \sum_{m=0}^{\infty} \frac{(-1)^m}{(2m)!} x^{2m} \int_0^{\pi/2} (\cos\theta)^{2\nu}(\sin\theta)^{2m}\,d\theta.$$

A reversal of the order of summation and integration gives

$$\Gamma\left(\nu + \tfrac{1}{2}\right) J_\nu(x) = \frac{2}{\sqrt{\pi}}\left(\frac{x}{2}\right)^\nu \int_0^{\pi/2} (\cos\theta)^{2\nu} \sum_{m=0}^{\infty} \frac{(-1)^m}{(2m)!} (x\sin\theta)^{2m}\,d\theta.$$

We recognize the sum as the cosine series for $x\sin\theta$. Thus,

$$J_\nu(x) = \frac{2}{\sqrt{\pi}\,\Gamma\left(\nu + \tfrac{1}{2}\right)} \left(\frac{x}{2}\right)^\nu \int_0^{\pi/2} \cos(x\sin\theta)(\cos\theta)^{2\nu}\,d\theta, \qquad (4.29)$$

which is *Poisson's integral*. A slight variation of this formula will be particularly useful in Chapter 12 in obtaining an important expansion of a plane wave. Other useful integral representations of $J_\nu(x)$ are derived in Chapter 10.

4.1.6   ZEROS OF BESSEL FUNCTIONS

Typically a Bessel function oscillates with diminishing amplitude as a function of its argument all the way out to infinity. This is illustrated in Figure 4.2 for several Bessel functions (of the first kind) of integral order.[15] Thus, the Bessel function of order ν has an infinite number of zeros. We denote the position of the nth zero of the νth order Bessel function by $x_{\nu n}$, that is

$$J_\nu(x_{\nu n}) = 0, \qquad n = 1, 2, 3, \dots .$$

4.1.7   BACK TO THE WAVE GUIDE

Return now to the problem of the cylindrical wave guide. Since the electromagnetic field must remain finite on the axis of the wave guide, we must discard the solution to Eq. (4.10) irregular at $\rho = 0$. According to Eq. (4.8), the field for TM modes is then

$$E_z(\rho, \theta) = J_n(\gamma\rho)\left(Ae^{in\theta} + Be^{-in\theta}\right),$$

where A and B are constants. With the inner radius of the wave guide equal to R the boundary condition is

$$E_z(\rho, \theta)\big|_{\rho=R} = 0,$$

---

[15]These curves were generated by numerical integration of *Bessel's integral*. See Eq. (9.16).

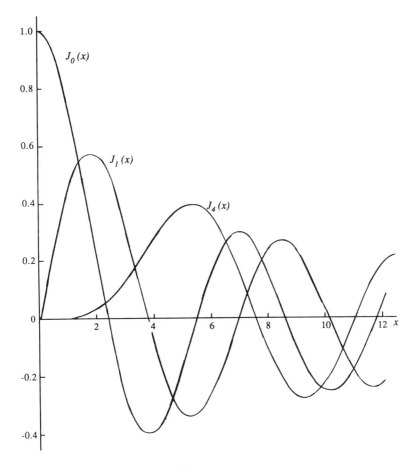

Figure 4.2

which holds for arbitrary θ. Thus,

$$J_n(\gamma R) = 0,$$

with

$$\gamma R = x_{nm}, \qquad m = 1, 2, 3, \ldots .$$

When the wave guide is excited by frequency ω, we see from Eq. (4.6) that only those wavelengths $\lambda = 2\pi/k$ that satisfy the condition

$$\frac{2\pi}{\lambda_{nm}} = k_{nm} = \sqrt{\mu\varepsilon\omega^2 - \frac{x_{nm}^2}{R^2}}$$

will propagate in the guide. All others are attenuated. Since the wavelength must be real, only those zeros of the Bessel function for which

$$x_{nm} < R\omega\sqrt{\mu\varepsilon}$$

correspond to modes that will propagate.

4.2   The Vibrating Membrane

Classical mechanics provides another example of how Bessel functions arise
in the description of the behavior of a physical system. A circular, elastic
membrane of radius b is stretched and clamped around the rim. When the
membrane is displaced from equilbrium and released, it vibrates due to
elastic restoring forces.[16] The displacement u of the membrane is described
by the two-dimensional wave equation[17]

$$\nabla^2 u(\rho,\theta,t) = \frac{1}{v^2}\frac{\partial^2}{\partial t^2}\,u(\rho,\theta,t), \tag{4.30}$$

where v is the propagation velocity of the wave. Again, we use the polar
coordinates (ρ,θ) of Figure 4.1.

Because of the separability of the space and time operators, the solutions
to Eq. (4.30) may be written as products of the type

$$u(\rho,\theta,t) = w(\rho,\theta)f(t).$$

On substituting this solution into Eq. (4.30) we find that f and w satisfy,
respectively, the differential equations,

$$\left(\frac{d^2}{dt^2} + k^2 v^2\right) f(t) = 0 \tag{4.31}$$

and

$$(\nabla^2 + k^2)w(\rho,\theta) = 0. \tag{4.32}$$

The parameter k is the separation constant.

The general solution to Eq. (4.31) is[18]

$$f(t) = A\cos\omega t + B\sin\omega t$$

where $\omega = kv$.

If we write the Laplacian ∇^2 in polar coordinates, Eq. (4.32) reads

$$\left[\frac{1}{\rho}\frac{\partial}{\partial\rho}\left(\rho\frac{\partial}{\partial\rho}\right) + \frac{1}{\rho^2}\frac{\partial^2}{\partial\theta^2} + k^2\right]w(\rho,\theta) = 0.$$

---

[16] An example is the vibration of a drumhead when it is struck.

[17] For a derivation, see A.L. Fetter and J.D. Walecka, *Theoretical Mechanics of
Particles and Continua*, McGraw-Hill, New York, 1980, p. 273.

[18] *Cf.* Eq. (1.8).

The operators are separable here, so again we have product solutions,

$$w(\rho, \theta) = R(\rho)S(\theta).$$

The functions R and S satisfy the differential equations,

$$\left(\frac{d^2}{d\theta^2} + \beta^2 \right) S(\theta) = 0, \tag{4.33}$$

and

$$\left(\rho \frac{d}{d\rho} \left(\rho \frac{d}{d\rho} \right) + (\rho^2 k^2 - \beta^2) \right) R(\rho) = 0. \tag{4.34}$$

As we saw earlier in this chapter, solutions to Eq. (4.33) are

$$S(\theta) = Ce^{i\beta\theta} + De^{-i\beta\theta}.$$

The solution must be single-valued. This requires that β be an integer, which we denote by $\beta = m$.

By defining a new, dimensionless variable $x = \rho k$, Eq. (4.34) becomes

$$\left[x^2 \frac{d^2}{dx^2} + x \frac{d}{dx} + (x^2 - m^2) \right] R(x) = 0,$$

which is Bessel's equation. The general solution is

$$R(\rho) = F_m J_m(k\rho) + G_m N_m(k\rho).$$

Since the membrane cannot experience an infinite displacement at its center ($\rho = 0$), we must discard the Neumann function by setting $G_m = 0$.

The differential equation (Eq. 4.30) is second order in time. To obtain the complete solution we must know the initial displacement and initial speed of the membrane. For example,[19]

$$u(\rho, \theta, t)\big|_{t=0} = f(\rho, \theta) \tag{4.35}$$

and

$$\frac{\partial}{\partial t} u(\rho, \theta, t)\big|_{t=0} = g(\rho, \theta). \tag{4.36}$$

We also have a boundary condition, namely that the membrane is fastened at the rim. This means that the displacement at $\rho = b$ is zero for all values of θ. Therefore,

$$J_m(kb) = 0.$$

Thus, the radial functions that satisfy the boundary condition are

$$J_m \left(x_{mn} \frac{\rho}{b} \right). \tag{4.37}$$

---

[19]These are the *initial conditions*.

Solutions to Eq. (4.32) are

$$w_{mn}(\rho, \theta) = J_m\left(x_{mn}\frac{\rho}{b}\right)e^{im\theta}. \tag{4.38}$$

We now have all of the parts of the solution to Eq. (4.30). In Chapter 12 we put them together to obtain the complete solution to the vibrating membrane problem. There we find it useful to have an expression for the normalization integral

$$I_{\nu n} = \int_0^b \left[J_\nu\left(x_{\nu n}\frac{\rho}{b}\right)\right]^2 \rho\, d\rho.$$

We change the variable of integration to $z = x_{\nu n}\rho/b$ and then integrate by parts. This yields

$$I_{\nu n} = \frac{b^2}{x_{\nu n}^2}\left[\tfrac{1}{2}z^2 J_\nu^2(z)\Big|_0^{x_{\nu n}} - \int_0^{x_{\nu n}} J_\nu'(z)J_\nu(z)z^2\, dz\right].$$

The first term in the brackets vanishes at both limits. In the second term we substitute for $J_\nu(z)z^2$ from Bessel's equation to obtain

$$I_{\nu n} = -\frac{b^2}{x_{\nu n}^2}\int_0^{x_{\nu n}} J_\nu'(z)\left[\nu^2 J_\nu(z) - z^2 J_\nu''(z) - z J_\nu'(z)\right]dz$$

$$= -\frac{b^2}{2x_{\nu n}^2}\int_0^{x_{\nu n}} \frac{d}{dz}\left[\nu^2 J_\nu^2(z) - [zJ_\nu'(z)]^2\right]dz$$

$$= -\frac{b^2}{2x_{\nu n}^2}\left[\nu^2 J_\nu^2(z) - [zJ_\nu'(z)]^2\right]\Big|_0^{x_{\nu n}}.$$

At both limits $\nu J_\nu(z) = 0$ for $\nu \geq 0$, whereas $zJ_\nu'(z) = 0$ at the lower limit. Thus,

$$I_{\nu n} = \frac{b^2}{2}\left[J_\nu'(x_{\nu n})\right]^2.$$

On substituting for $J_\nu'(x_{\nu n})$ from either of the recursion formulas, Eq. (4.22) or Eq. (4.23), and recognizing that $J_\nu(x_{\nu n}) = 0$, we get

$$\int_0^b \left[J_\nu\left(x_{\nu n}\frac{\rho}{b}\right)\right]^2 \rho\, d\rho = \frac{b^2}{2}J_{\nu\pm1}^2(x_{\nu n}). \tag{4.39}$$

EXERCISES

1. Use the method of separation of variables to obtain the general solution to the partial differential equation

$$\frac{\partial^2}{\partial x^2}f(x,t) - \gamma^2\frac{\partial}{\partial t}f(x,t) = 0,$$

where γ is real. Denote the separation constant by $-k^2$ and assume that physical considerations require that k be real.

2. Solve Bessel's equation

$$x^2 \frac{d^2 u}{dx^2} + x \frac{du}{dx} + (x^2 - \nu^2)u = 0$$

by the series method. That is, assume the solution is

$$u(x) = \sum_{k=0}^{\infty} c_k x^{k+s}.$$

Show that for $k > 1$ the coefficients c_k are related by the recursion formula,

$$[(k+s)^2 - \nu^2]c_k + c_{k-2} = 0.$$

For $\nu^2 \neq \frac{1}{4}$ verify that, if $c_0 \neq 0$, then $c_1 = 0$ and *vice versa*. Show that the solutions obtained here can be expressed in terms of generalized hypergeometric functions.[20]

3. Write Bessel's equation in the standard form of Eq. (2.24) and show that the singularity at $x = 0$ is regular and the one at $x = \infty$ is irregular.

4. With the change of dependent variable $w(x) = e^{bx} f(x)$ show that the differential equation

$$xw''(x) + 2\lambda w'(x) + xw(x) = 0$$

can be transformed into the confluent hypergeometric equation. Write down the general solution in terms of confluent hypergeometric functions. Show that the solution can also be expressed in terms of Bessel functions.

5. Use Eqs. (4.21) and (4.24) to prove that

$$\frac{d}{dt}\left[t^{\frac{1}{2}\nu} J_\nu(2\sqrt{t})\right] = t^{\frac{1}{2}(\nu-1)} J_{\nu-1}(2\sqrt{t}).$$

6. Consider the differential equation[21]

$$\frac{d^2}{dx^2} u(x) + \lambda x^s u(x) = 0$$

where λ is a constant and s is a real positive number. This equation can be transformed into Bessel's equation by changes of dependent and independent variables according to

$$u(x) = x^p f(x) \quad \text{and} \quad z = bx^q.$$

---

[20] *Cf.* Exercise 1.5.

[21] Solutions to this equation are useful in constructing the connection formulas of the WKB approximation. See, for example, L.I. Schiff, *Quantum Mechanics*, McGraw-Hill, New York, 1955, p. 187.

Find the values of the constants b, p, and q required to effect the transformation and write down explicitly the differential equation for f as a function of z. Show that the general solution to the original equation is

$$u(x) = A\sqrt{x}\,J_\nu\left(2\nu\sqrt{\lambda}x^{1/2\nu}\right) + B\sqrt{x}N_\nu\left(2\nu\sqrt{\lambda}x^{1/2\nu}\right)$$

where $\nu = (s+2)^{-1}$. The Schrödinger equation for a particle moving under the influence of a constant force can be transformed into this equation with $s = 1$.[22]

7. Use the definition of the Neumann function $N_\nu(x)$ given in Eq. (4.26) to show that

$$\frac{d}{dx}N_\nu(x) = N_{\nu-1}(x) - \frac{\nu}{x}N_\nu(x).$$

Show that other recursion formulas for $N_\nu(x)$ analogous to Eqs. (4.21), (4.22), and (4.24) also follow from the definition of $N_\nu(x)$. These formulas hold for any linear combination of $J_\nu(x)$ and $N_\nu(x)$ including the Hankel functions $H_\nu^{(1)}(x)$ and $H_\nu^{(2)}(x)$.

8. Show that *any* function $B_\nu(x)$ that satisfies the basic recursion formulas

$$\frac{d}{dx}B_\nu(x) = B_{\nu-1}(x) - \frac{\nu}{x}B_\nu(x)$$

and

$$\frac{d}{dx}B_\nu(x) = -B_{\nu+1}(x) + \frac{\nu}{x}B_\nu(x)$$

also satisfies Bessel's equation.

9. By a change of dependent variable Bessel's equation may be transformed into the equation

$$\frac{d^2}{dx^2}y(x) + \left(1 - \frac{\left(\nu^2 - \frac{1}{4}\right)}{x^2}\right)y(x) = 0.$$

Write the general solution explicitly in terms of Bessel functions and Neumann functions of order ν. By inspection of this equation what kind of solutions would you expect for $\nu = \pm\frac{1}{2}$? Is this consistent with the solutions in terms of Bessel functions? Explain.

---

[22] See Exercise 10.12.

5

The Central Force Problem in Quantum Mechanics

One of the most important applications of quantum mechanics is to describe the behavior of a particle moving under the influence of a central force field. Not only is the problem of pedagogical interest in its own right (Coulomb field, isotropic harmonic oscillator, etc.), but it is also the starting point for perturbation calculations for more realistic situations in atomic and nuclear physics. Angular momentum eigenfunctions have a wide application at all levels of quantum physics. Some examples will follow.

5.1 Three-Dimensional Schrödinger Equation

In three dimensions the time-independent Schrödinger equation for a particle of mass m is

$$\left[\frac{-\hbar^2}{2m}\nabla^2 + V(\mathbf{r})\right] u(\mathbf{r}) = Eu(\mathbf{r}). \tag{5.1}$$

This is a partial differential equation in three independent variables—the three space coordinates. For a central force the potential energy $V(\mathbf{r})$ at a point $\mathbf{r}$ depends only on the distance from the origin and not on the direction. If we represent the position vector $\mathbf{r}$ in terms of the spherical polar coordinates,

$$\mathbf{r} = (r, \theta, \phi) \tag{5.2}$$

defined in Figure 5.1, then $V(\mathbf{r}) = V(r)$ for a central force and Eq. (5.1) can be solved by the method of separation of variables. In this system the differential operator ∇^2 is

$$\nabla^2 = \frac{1}{r^2}\frac{\partial}{\partial r}\left(r^2\frac{\partial}{\partial r}\right) + \frac{1}{r^2 \sin\theta}\frac{\partial}{\partial \theta}\left(\sin\theta\frac{\partial}{\partial \theta}\right) + \frac{1}{r^2 \sin^2\theta}\frac{\partial^2}{\partial \phi^2}. \tag{5.3}$$

This substitution for ∇^2 in Eq. (5.1) with $V(\mathbf{r}) = V(r)$ followed by a rearrangement of terms yields the equation,

$$\left\{\left[\frac{\partial}{\partial r}\left(r^2\frac{\partial}{\partial r}\right) + \frac{2mr^2}{\hbar^2}(E - V(r))\right] + \left[\frac{1}{\sin\theta}\frac{\partial}{\partial \theta}\left(\sin\theta\frac{\partial}{\partial \theta}\right)\right.\right.$$
$$\left.\left. + \frac{1}{\sin^2\theta}\frac{\partial^2}{\partial \phi^2}\right]\right\}u(\mathbf{r}) = 0. \tag{5.4}$$

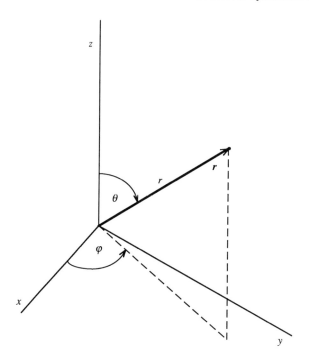

Figure 5.1

We see that the operator in the curly brackets separates into two parts—one depending only on the radial coordinate r and the other depending only on the angular coordinates θ and ϕ. The angular part is seen to be proportional to the quantum operator corresponding to the orbital angular momentum[1] of the particle,

$$L^2 = -\hbar^2 \left[\frac{1}{\sin\theta} \frac{\partial}{\partial\theta} \left(\sin\theta \frac{\partial}{\partial\theta} \right) + \frac{1}{\sin^2\theta} \frac{\partial^2}{\partial\phi^2} \right], \qquad (5.5)$$

with the projection of **L** onto the z axis given by

$$L_z = -i\hbar \frac{\partial}{\partial\phi}. \qquad (5.6)$$

According to the method of separation of variables we write the solution to Eq. (5.4) as a product

$$u(\mathbf{r}) = R(r)Y(\theta, \phi), \qquad (5.7)$$

---

[1]D.S. Saxon, *Elementary Quantum Mechanics*, Holden Day, San Francisco, 1968, p. 272. E. Merzbacher, *op. cit.*, p. 173.

which we substitute into Eq. (5.4). After some rewriting we obtain

$$\frac{1}{R}\left[\frac{d}{dr}\left(r^2\frac{d}{dr}\right) + \frac{2mr^2}{\hbar^2}(E - V(r))\right]R$$

$$= \frac{-1}{Y}\left[\frac{1}{\sin\theta}\frac{\partial}{\partial\theta}\left(\sin\theta\frac{\partial}{\partial\theta}\right) + \frac{1}{\sin^2\theta}\frac{\partial^2}{\partial\phi^2}\right]Y. \qquad (5.8)$$

Clearly, the left-hand side depends only on r (which justifies replacing the partial derivative with the ordinary derivative), whereas the right-hand side depends only on the angular variables. Since r and the angles vary independently, the two sides of Eq. (5.8) must be equal to a constant. If we denote this constant by λ and take the right-hand side of the equation, Eq. (5.8) may be written as

$$\left\{\left[\sin\theta\frac{\partial}{\partial\theta}\left(\sin\theta\frac{\partial}{\partial\theta}\right) + \lambda\sin^2\theta\right] + \frac{\partial^2}{\partial\phi^2}\right\}Y(\theta,\phi) = 0. \qquad (5.9)$$

From Eq. (5.5) it follows that the angular momentum eigenvalue equation is

$$L^2Y(\theta,\phi) = \lambda\hbar^2Y(\theta,\phi).$$

Presently, we shall show that the orbital angular momentum is quantized.

The operator in curly brackets in Eq. (5.9) is separable into a θ-dependent part and a ϕ-dependent part. Therefore, we make the substitution

$$Y(\theta,\phi) = f(\theta)g(\phi),$$

to obtain

$$\frac{1}{f(\theta)}\left[\sin\theta\frac{d}{d\theta}\left(\sin\theta\frac{d}{d\theta}\right) + \lambda\sin^2\theta\right]f(\theta) = \frac{-1}{g(\phi)}\frac{d^2}{d\phi^2}g(\phi). \qquad (5.10)$$

The left-hand side of this equation depends only on θ, whereas the right-hand side depends only on ϕ. Since θ and ϕ vary independently, the two sides must be equal to the same constant, say β. For the right-hand side this gives

$$\frac{d^2}{d\phi^2}g + \beta g = 0,$$

which has solutions

$$g(\phi) = e^{\pm i\sqrt{\beta}\phi}. \qquad (5.11)$$

On physical grounds we require that $u(\mathbf{r})$ be single-valued. This means that at a given point in space $u(\mathbf{r})$ has a unique value. Since (r,θ,ϕ) and $(r,\theta,\phi + 2\pi)$ represent the same location in space, we must have

$$g(\phi + 2\pi) = g(\phi).$$

From Eq. (5.11) we see that this holds only if $\sqrt{\beta}$ is zero or an integer. We set $\beta = m^2$ and write

$$g(\phi) = e^{im\phi}, \qquad m = 0, \pm 1, \pm 2, \ldots.$$

According to Eq. (5.6) we have the eigenvalue equation

$$L_z g(\phi) = m\hbar g(\phi).$$

Thus, the physical requirement of single-valuedness of the solution of the Schrödinger equation results in a separation constant, which is discrete. The physical significance of this is that the direction of the angular momentum vector is *quantized*. It is restricted to certain allowed orientations in space. From the left-hand side of Eq. (5.10) we obtain after some differentiation and rearrangement the equation,

$$\sin^2 \theta \frac{d^2 f}{d\theta^2} + \sin \theta \cos \theta \frac{df}{d\theta} + (\lambda \sin^2 \theta - m^2) f = 0.$$

5.2   Legendre's Equation

To cast this latter equation into standard form (Eq. 2.24) with rational functions for the coefficients in each term, we make the change of variable $x = \cos \theta$ and a change in notation which replaces $f(\theta)$ by $f(x)$. The result is

$$(1 - x^2) \frac{d^2}{dx^2} f(x) - 2x \frac{d}{dx} f(x) + \left(\lambda - \frac{m^2}{1 - x^2} \right) f(x) = 0. \qquad (5.12)$$

This is a second-order differential equation in x with three regular singular points (at $x = \infty$ and $x = \pm 1$). With three regular singularities we look for suitable changes of variables that cast it into the form of the hypergeometric equation.

We begin with the dependent variable. Let us make the substitution

$$f(x) = v(x) w(x)$$

in Eq. (5.12) and obtain a differential equation for $w(x)$. We get

$$(1 - x^2) v w''(x) + \left[2(1 - x^2) v' - 2xv \right] w'(x)$$

$$+ \left[(\lambda - (1 - x^2)^{-1} m^2) v + (1 - x^2) v'' - 2xv' \right] w(x) = 0. \qquad (5.13)$$

On comparing the second-order term and the zero-order term we see that the appropriate form of Eq. (2.26) requires that the factor

$$(1 - x^2) v'' - 2xv' + \left(\lambda - m^2 (1 - x^2)^{-1} \right) v \qquad (5.14)$$

be proportional to v. We can absorb any overall constant factor into $w(x)$. Therefore, we set $v = (1 - x^2)^a$. On substituting this into Eq. (5.14) we find that we must have $4a^2 = m^2$. So,

$$v = (1 - x^2)^{\pm \frac{1}{2}|m|}.$$

Thus, Eq. (5.13) for $w(x)$ becomes

$$(1 - x^2)w''(x) - 2(1 \pm |m|)xw'(x) - (m^2 \pm |m| - \lambda)w(x) = 0. \quad (5.15)$$

Now, to move the singularities to the standard positions, we change the independent variable so that $(1 - x^2)$ is replaced by $z(1 - z)$. Set

$$1 - x^2 = \alpha z(1 - z).$$

On solving for z in terms of x and choosing $\alpha = 4$, we find that

$$z = \tfrac{1}{2}(1 - x) \quad (5.16)$$

will satisfy our requirements. With this variable change Eq. (5.15) becomes

$$z(1 - z)\frac{d^2}{dz^2}w(z) + \left[1 \pm |m| - 2(1 \pm |m|)z\right]\frac{d}{dz}w(z)$$

$$- [m^2 \pm |m| - \lambda]w(z) = 0, \quad (5.17)$$

which is exactly the form of the hypergeometric equation (Eq. 2.26). The general solution is given in Eq. (2.36). We note that, when $\theta = 0$, $z = 0$. Since such points in space are physically allowed, we must discard the solution that is singular at $z = 0$. This means setting $B = 0$ in Eq. (2.36). Thus, the solution to Eq. (5.17) that is acceptable on physical grounds is

$$w(z) = {}_2F_1(a, b; c; z),$$

where $a + b = \pm 2|m| + 1$, $ab = m^2 \pm |m| - \lambda$, and $c = \pm |m| + 1$. The function ${}_2F_1(a, b; c; z)$ is invariant under interchange of a and b. We solve for a and b and choose b to be the larger of the two. This gives

$$w(z) = {}_2F_1\left(\pm|m| + \tfrac{1}{2}(1 - \sqrt{4\lambda + 1}), \pm|m| + \tfrac{1}{2}(1 + \sqrt{4\lambda + 1}); \pm|m| + 1; z\right).$$
$$(5.18)$$

To find the right choice of sign here we examine the behavior of the complete solution at the physically accessible point $z = 0$ ($x = 1$ or $\theta = 0$). The complete solution to Eq. (5.12) is

$$f(x) = A(1 - x^2)^{\pm \frac{1}{2}|m|}\left[1 + \sum_{k=1}^{\infty}\frac{(a)_k(b)_k}{k!(c)_k}\left(\tfrac{1}{2} - \tfrac{1}{2}x\right)^k\right],$$

with a, b, and c defined above. As x approaches 1 the quantity in the square brackets also approaches unity, whereas $(1 - x^2)$ approaches zero. Thus, for

$f(x)$ to be well behaved at $x = 1$ we must choose the upper sign. The complete solution to Eq. (5.12) which meets the physical requirements is

$$f(x) = A(1 - x^2)^{\frac{1}{2}|m|} {}_2F_1\left(|m| + \tfrac{1}{2}(1 - \sqrt{4\lambda + 1}),\right.$$
$$\left. |m| + \tfrac{1}{2}(1 + \sqrt{4\lambda + 1}); |m| + 1; \tfrac{1}{2} - \tfrac{1}{2}x\right).$$

Now let us examine the behavior of ${}_2F_1(a, b; c; z)$ for $z = 1$ ($x = -1$ or $\theta = \pi$), which is also a value permitted by the physics. We have

$$_2F_1(a, b; c; 1) = \sum_{k=0}^{\infty} \frac{(a)_k (b)_k}{k! (c)_k}.$$

Clearly, this series diverges if the $k + 1$ term is not smaller than the kth term, that is if

$$\frac{(a)_{k+1}(b)_{k+1}}{(k+1)!(c)_{k+1}} \geq \frac{(a)_k(b)_k}{k!(c)_k},$$

or $(a + k)(b + k) \geq (k + 1)(c + k)$. Expansion of the two sides leads to $ab + (a + b)k \geq c + ck$. On comparing terms for large k we see that the series diverges for $x = -1$, if $a + b \geq c$. Since $|m| \geq 0$, this condition is fulfilled ($a + b = 2|m| + 1$ and $c = |m| + 1$). So, to obtain a physically acceptable solution, the series must terminate. Either a or b must be a negative integer or zero. Now, $a + b = 2|m| + 1$ is an odd integer, so $a \neq b$. By choice $b > a$. Also if either a or b is an integer, then both are integers. If b is a negative integer, then so is a, since $b > a$. Therefore, a will cause the series to terminate before b does. Therefore, it is sufficient to require that a be a negative integer or zero,

$$a = |m| + \tfrac{1}{2}(1 - \sqrt{4\lambda + 1}) = -n, \qquad n = 0, 1, 2, \dots,$$
$$b = n + 2|m| + 1.$$

We solve for λ, the separation constant,

$$\lambda = (n + |m|)(n + |m| + 1).$$

Note that $n + |m|$ is a nonnegative integer. If we denote this integer by $l = n + |m|$, then the separation constant is

$$\lambda = l(l + 1), \qquad l = 0, 1, 2, \dots.$$

The physical implication of the discreteness of the separation constant (l can have only *integer* values) is that the *magnitude* of the orbital angular momentum is quantized as is its orientation in space. The length of the angular momentum vector can take on only certain allowed values. Now, $l - |m| = n \geq 0$. Thus, $l \geq |m|$ and m lies in the range[2]

---

[2]This is simply a reflection of the fact that a projection of the angular momentum vector onto some axis cannot be greater than the magnitude of the vector itself.

$$-l \leq m \leq l.$$

In terms of these new parameters the required solution of Eq. (5.12) is

$$f_{lm}(x) = A_{lm}(1-x^2)^{\frac{1}{2}|m|} \, {}_2F_1\left(-l+|m|, l+|m|+1; |m|+1; \tfrac{1}{2}-\tfrac{1}{2}x\right). \quad (5.19)$$

With $m = 0$ and $\lambda = l(l+1)$ Eq. (5.12) becomes

$$(1-x^2)f''(x) - 2xf'(x) + l(l+1)f(x) = 0,$$

which is *Legendre's equation*.

5.3   Legendre Polynomials and Associated Legendre Functions

From Eq. (5.19) we see that a solution to Legendre's equation above is

$$P_l(x) = {}_2F_1\left(-l, l+1; 1; \tfrac{1}{2}(1-x)\right), \quad (5.20)$$

which is a polynomial. This function is called the *Legendre polynomial* of order l. Let us differentiate this function p times,

$$\frac{d^p}{dx^p}P_l(x) = (-1)^p \sum_{k=p}^{\infty} \frac{(-l)_k(l+1)_k}{2^k k!(1)_k}(k-p+1)_p(1-x)^{k-p}.$$

We define a new summation index by replacing $k-p$ by k. Then, with some manipulation of the Pochhammer symbols and using the identity

$$(a)_{k+p} = (a)_p(a+p)_k$$

we get

$$\frac{d^p}{dx^p}P_l(x) = (-1)^p \frac{(-l)_p(l+1)_p}{2^p p!} \sum_{k=0}^{\infty} \frac{(-l+p)_k(l+p+1)_k}{2^k k!(p+1)_k}(1-x)^k$$

$$= (-1)^p \frac{(-l)_p(l+1)_p}{2^p p!} \, {}_2F_1\left(-l+p, l+p+1; p+1; \tfrac{1}{2}(1-x)\right). \quad (5.21)$$

From this result we see by comparison with Eq. (5.19) that our solution to Eq. (5.12) can be written as

$$P_l^m(x) = (1-x^2)^{\frac{1}{2}|m|}\frac{d^{|m|}}{dx^{|m|}}P_l(x). \quad (5.22)$$

This function is known as the *associated Legendre function*. From Eqs. (5.21) and (5.22) we find that it is related to the hypergeometric function by

$$P_l^m(x) = \frac{(l+|m|)!(1-x^2)^{\frac{1}{2}|m|}}{2^{|m|}|m|!(l-|m|)!}$$

$$\times \, _2F_1\left(-l+|m|, l+|m|+1; |m|+1; \tfrac{1}{2}(1-x)\right). \tag{5.23}$$

The Legendre polynomials are *orthogonal*[3] functions, by which we mean that they have the property,

$$\int_{-1}^{1} P_l(x)P_{l'}(x)dx = 0, \qquad l' \neq l. \tag{5.24}$$

To prove this write Legendre's equation for two *different* Legendre polynomials,

$$(1-x^2)P_{l'}''(x) - 2xP_{l'}'(x) + l'(l'+1)P_{l'}(x) = 0,$$
$$(1-x^2)P_l''(x) - 2xP_l'(x) + l(l+1)P_l(x) = 0,$$

where $l' \neq l$. Multiply the first of these equations by $P_l(x)$, the second by $P_{l'}(x)$, and substract the second from the first to obtain

$$(1-x^2)\left[P_lP_{l'}'' - P_l''P_{l'}\right] - 2x\left[P_lP_{l'}' - P_l'P_{l'}\right]$$
$$+ \left[l'(l'+1) - l(l+1)\right]P_lP_{l'} = 0.$$

On combining the first two terms and rearranging we have

$$\frac{d}{dx}\left\{(1-x^2)\left[P_lP_{l'}' - P_l'P_{l'}\right]\right\} = \left[l(l+1) - l'(l'+1)\right]P_lP_{l'}.$$

If we integrate both sides from -1 to $+1$ and recognize that the integrand on the left-hand side is an exact differential, we get

$$(1-x^2)\left[P_l(x)P_{l'}'(x) - P_l'(x)P_{l'}(x)\right]\Big|_{-1}^{+1}$$
$$= \left[l(l+1) - l'(l'+1)\right]\int_{-1}^{1} P_l(x)P_{l'}(x)dx.$$

The left-hand side is zero, since the factor $(1-x^2)$ vanishes at both limits. Thus, Eq. (5.24) is established. In Chapter 12 we obtain a more general relation of this type for the associated Legendre functions, $P_l^m(x)$.

---

[3]The properties of orthogonal functions will be discussed more fully in Chapter 12.

5.4 Eigenfunctions for a Particle in a Central Field

We are now in a position to write the solution to Schrödinger's equation for any central force field. According to Eq. (5.7)

$$u(\mathbf{r}) = R(r)Y_l^m(\theta, \phi), \tag{5.25}$$

where

$$Y_l^m(\theta, \phi) = C_{lm}e^{im\phi}P_l^m(\cos\theta). \tag{5.26}$$

The constant C_{lm} is a normalization factor, which we evaluate in Chapter 12. The radial function $R(r)$ depends on the quantum number l, as well as on the particular central force considered, as discussed in the next chapter.

EXERCISES

1. For $m = 0$ Eq. (5.12) reduces to

$$(1 - x^2)f''(x) - 2xf'(x) + \lambda f(x) = 0.$$

Assume that this equation has series solutions of the form

$$f(x) = \sum_{k=0}^{\infty} a_k x^{k+s}.$$

Substitute this series into the differential equation, determine the possible values of s, and obtain a recursion formula for the coefficients a_k. To obtain a polynomial solution we cannot have both a_0 and a_1 different from zero at the same time. Why not? Explain. Consider the case for $s = 0$ and show that polynomial solutions require that

$$\lambda = n(n+1), \qquad n = 0, 1, 2, \dots.$$

Use the identity

$$k(k+1) - n(n+1) = (k-n)(k+n+1)$$

to show that for even n the solution can be written as

$$f_n(x) = a_0(-1)^{n/2}\frac{\left[\left(\frac{n}{2}\right)!\right]^2}{n!}\sum_{k=0}^{n/2}\frac{(-1)^k(2n-2k)!}{k!(n-2k)!(n-k)!}x^{n-2k}$$

which is a polynomial in even powers of x.

2. Find the solution to Legendre's equation in the preceding problem when n is odd. Show that by choosing a_0 and a_1 according to

$$a_0 = \frac{(-1)^{n/2} n!}{2^n \left[\left(\frac{n}{2} \right)! \right]^2}, \qquad n = \text{even},$$

$$a_1 = \frac{(-1)^{(n-1)/2} (n+1)!}{2^n \left(\frac{n+1}{2} \right)! \left(\frac{n-1}{2} \right)!}, \qquad n = \text{odd},$$

a solution to Legendre's equation can be written as[4]

$$f_n(x) = \sum_{k=0}^{[n/2]} \frac{(-1)^k (2n - 2k)!}{2^n k! (n - 2k)! (n - k)!} x^{n-2k}$$

where n is either even or odd.

3. The differential equation

$$(1 - x^2) \frac{d^2 y}{dx^2} - x \frac{dy}{dx} + n^2 y = 0$$

has three singularities. Find the values of x that are the singular points. Use the criteria in Section 2.3.3 to show that they are all regular singularities. Find a suitable change of independent variable that puts this equation in the form of the hypergeometric equation. Write the general solution $y(x)$ explicitly in terms of hypergeometric functions. Show that if n is any integer, one of the solutions is a polynomial.[5]

4. Obtain the general solution to the differential equation

$$(1 - x^2) \frac{d^2 y}{dx^2} - 3x \frac{dy}{dx} + n(n + 2)y = 0$$

in terms of hypergeometric functions. Show that if n is an integer $(n \neq -1)$, then one of the solutions is a polynomial.[6]

5. The potential energy of a particle of mass m constrained to move along the x axis is given by[7]

$$V(x) = \frac{-V_0}{\cosh^2(x/b)},$$

---

[4]In Chapter 9, we show that this function is just the Legendre polynomial $P_n(x)$ defined by Eq. (5.20).

[5]These are known as *Chebyshev polynomials type* 1. They are very useful in numerical analysis.

[6]These are *Chebyshev polynomials of type* 2.

[7]D. ter Haar, *Selected Problems in Quantum Mechanics*, Academic Press, New York, 1975, p. 5. Used with permission of Pion Limited, London. For definitions of the hyperbolic functions $\sinh y$ and $\cosh y$, see p. 147.

where V_0 and b are constants. This function is shown graphically in Figure 5.2. Schrödinger's equation for the particle can be written as

$$\frac{d^2}{dx^2} u(x) + \frac{2m}{\hbar^2} \left(E + \frac{V_0}{\cosh^2(x/b)} \right) u(x) = 0.$$

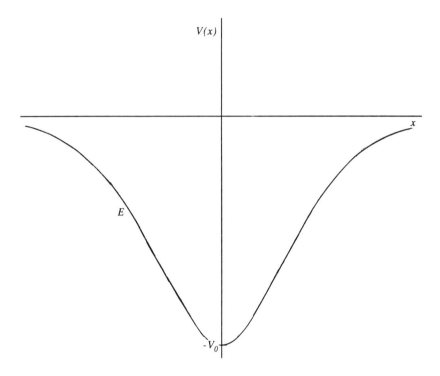

Figure 5.2

In seeking to transform this equation into the hypergeometric equation we see from Eq. (2.26) that the coefficient of the dependent variable in the zero-order term must be constant. This leads us to try in the equation above, the substitution

$$u(x) = \left(\cosh(x/b) \right)^p g(x).$$

Find the value of p such that the differential equation for $g(x)$ reads

$$\frac{d^2}{dx^2} g(x) + \mu \tanh\left(\frac{x}{b} \right) \frac{d}{dx} g(x) + \lambda g(x) = 0$$

where μ and λ are constants.

Next, we look for a change of independent variable which will cast the first-order term into the appropriate form for the hypergeometric equation. Since the coefficient in the first-order term here is a hyperbolic function, let us try

$$z = \big(\cosh(x/b)\big)^q.$$

Find the value of q for which the differential equation for $g(x)$ may be written as

$$z(1 - z)\frac{d^2}{dz^2}g(z) + (\alpha - \beta z)\frac{d}{dz}g(z) + \gamma g(z) = 0$$

where α, β, and γ are constants. A further change of variable $y = 1 - z$ also yields the hypergeometric equation.[8] From these results write down the general solution to the Schrödinger equation explicitly in terms of hypergeometric functions. Invoke the boundary conditions at $x = 0$ at $x = \pm\infty$ to obtain the physically acceptable solutions for bound states $(E < 0)$. Show that the energy spectrum is discrete with energy eigenvalues

$$E_n = \frac{-\hbar^2}{2mb^2}\left[\frac{1}{2}\sqrt{\frac{8mV_0b^2}{\hbar^2} + 1} - \left(n + \frac{1}{2}\right)\right]^2, \qquad n = 0, 1, 2, \ldots,$$

and corresponding eigenfunctions

$$u_n(x) = A_n\left(\cosh\left(\frac{x}{b}\right)\right)^{p/2} {}_2F_1\left(\frac{-n}{2}, \frac{-n}{2} + \sigma; \frac{1}{2}; -\sinh^2\left(\frac{x}{b}\right)\right),$$
$$n = 0, 2, 4, \ldots,$$

$$u_n(x) = A_n\left(\cosh\left(\frac{x}{b}\right)\right)^{p/2}\sinh\left(\frac{x}{b}\right)$$
$$\times {}_2F_1\left(\frac{1-n}{2}, \frac{1-n}{2} + \sigma; \frac{3}{2}; -\sinh^2\left(\frac{x}{b}\right)\right), \qquad n = 1, 3, 5, \ldots,$$

where $\sigma = \sqrt{-2mEb^2/\hbar^2}$.

---

[8] See Exercise 2.18.

6

The Radial Equation for Central Force Fields

We saw in the previous chapter that for spherically symmetric force fields solutions to Schrödinger's equation are

$$u(\mathbf{r}) = R(r)Y(\theta, \phi),$$

where $Y(\theta, \phi)$ satisfies Eq. (5.9). The general solution can be written as an infinite sum of such products.[1] The radial dependence of $u(\mathbf{r})$ is contained in $R(r)$ which is described by the differential equation[2]

$$\frac{d}{dr}\left(r^2\frac{d}{dr}R(r)\right) + \left\{\frac{2mr^2}{\hbar^2}[E - V(r)] - l(l+1)\right\}R(r) = 0. \qquad (6.1)$$

The function $V(r)$ represents the potential energy of the particle in the central field.

Among the most frequently encountered central force fields in quantum mechanics are the square well, the harmonic oscillator, and the Coulomb field. For each of these Eq. (6.1) can be related to the confluent hypergeometric equation and the radial functions can be expressed in terms of the corresponding confluent hypergeometric functions.

6.1 Square Well

The potential energy function

$$V(r) = \begin{cases} -V_0, & r \le b, \\ \\ 0, & r > b, \end{cases} \qquad (6.2)$$

is shown in Figure 6.1a. For $r \le b$, Eq. (6.1) becomes

$$\frac{d^2}{dr^2}R(r) + \frac{2}{r}\frac{d}{dr}R(r) + \left[\frac{2m}{\hbar^2}(E + V_0) - \frac{l(l+1)}{r^2}\right]R(r) = 0. \qquad (6.3)$$

---

[1] We shall discuss this in more detail in Chapter 12.

[2] This equation follows from Eq. (5.8) with the separation constant $\lambda = l(l+1)$.

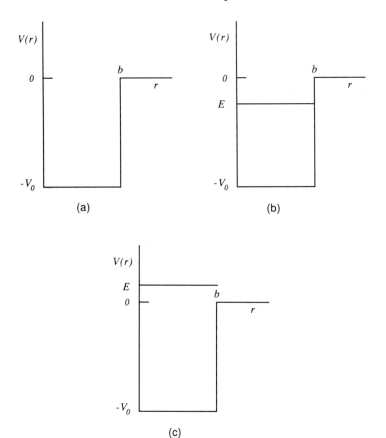

Figure 6.1

It is clear from Figures 6.1b and 6.1c that $-V_0 < E$ for both bound and unbound states. Therefore, we change to the dimensionless independent variable $\rho = \alpha r$ where

$$\alpha = \left[2m(E + V_0)/\hbar^2\right]^{1/2} \tag{6.4}$$

is real. This gives

$$\frac{d^2}{d\rho^2}R(\rho) + \frac{2}{\rho}\frac{d}{d\rho}R(\rho) + \left(1 - \frac{l(l+1)}{\rho^2}\right)R(\rho) = 0, \tag{6.5}$$

which *almost* has the form of Bessel's equation. We try to find a change of dependent variable

$$R(\rho) = v(\rho)w(\rho) \tag{6.6}$$

to transform Eq. (6.5) into Bessel's equation. By making this substitution for $R(\rho)$ in Eq. (6.5), we get after some rearrangement

$$v\rho^2 w'' + 2\rho(v + \rho v')w' + \left[\rho^2 v'' + 2\rho v' + \rho^2 v - l(l+1)v\right]w = 0. \quad (6.7)$$

From a comparison of the first- and second-order terms with the corresponding terms in Eq. (4.11) we see that we need

$$v + \rho v' = \tfrac{1}{2}v$$

from which we find that $v = \rho^{-1/2}$. This result inserted into Eq. (6.7) yields

$$\rho^2 w''(\rho) + \rho w'(\rho) + \left[\rho^2 - \left(l + \tfrac{1}{2}\right)^2\right]w(\rho) = 0 \quad (6.8)$$

which is Bessel's equation. Because $l + \tfrac{1}{2}$ is *not* an integer, the general solution of Eq. (6.5) is

$$R_l(\rho) = \rho^{-1/2}\left[A_l J_{l+\frac{1}{2}}(\rho) + B_l J_{-l-\frac{1}{2}}(\rho)\right].$$

We define the functions

$$j_l(\rho) = \sqrt{\frac{\pi}{2\rho}} J_{l+\frac{1}{2}}(\rho), \quad (6.9a)$$

$$n_l(\rho) = (-1)^{l+1}\sqrt{\frac{\pi}{2\rho}} J_{-l-\frac{1}{2}}(\rho). \quad (6.9b)$$

These functions are called, respectively, the *spherical Bessel function* and the *spherical Neumann function*. With a slight alteration of notation, we have for the general solution of Eq. (6.5)

$$R_l(r) = a_l j_l(\alpha r) + b_l n_l(\alpha r). \quad (6.10)$$

From the leading terms in the series expansions[3] of Eqs. (6.9),

$$j_l(x) = \frac{\sqrt{\pi}}{2}\left(\frac{x}{2}\right)^l\left[\frac{1}{\Gamma(l + \frac{3}{2})} + O(x^2)\right] \quad (6.11a)$$

$$n_l(x) = \frac{\sqrt{\pi}}{2}\left(\frac{-2}{x}\right)^{l+1}\left[\frac{1}{\Gamma(-l + \frac{1}{2})} + O(x^2)\right], \quad (6.11b)$$

we see that because $l \geq 0$, $n_l(x)$ diverges as $x \to 0$, whereas $j_l(x)$ does not. Thus, $j_l(x)$ is regular at $x = 0$ and $n_l(x)$ is irregular at that point. Equation (6.10) represents the particle for $0 \leq r \leq b$, therefore, we must take $b_l = 0$. Then,

$$R_l(r) = a_l j_l(\alpha r), \qquad r \leq b. \quad (6.12)$$

Outside the potential well ($r > b$) the potential energy of the particle is zero. For this region we can make a change of independent variable in Eq. (6.1) similar to that for $r \leq b$. In Eq. (6.3), we set $V_0 = 0$ and distinguish two cases.

---

[3]Refer to Eqs. (4.25).

6.1.1 BOUND STATES $(E < 0)$

This is the situation depicted in Figure 6.1b in which the corresponding classical particle is confined to the region $r \leq b$. In this case we take $\rho = i\beta r$ where

$$\beta = \left(2m|E|/\hbar^2\right)^{1/2} \tag{6.13}$$

is real. Again we obtain Eq. (6.5), which has solutions

$$R_l(r) = c_l j_l(i\beta r) + d_l n_l(i\beta r), \qquad r > b.$$

Because $r = 0$ is outside the range of validity of this solution, we cannot require that $d_l = 0$. In this region it is convenient to use the linear combinations,

$$h_l^{(1)}(\rho) = j_l(\rho) + i n_l(\rho), \tag{6.14a}$$

and

$$h_l^{(2)}(\rho) = j_l(\rho) - i n_l(\rho). \tag{6.14b}$$

The reason for this choice becomes clear presently.

The functions $h_l^{(1)}(\rho)$ and $h_l^{(2)}(\rho)$ are called the *spherical Hankel functions* of the *first* and *second kind*, respectively. With these definitions, the solutions to Eq. (6.1) for $r > b$ are

$$R_l(r) = c_l' h_l^{(1)}(i\beta r) + d_l' h_l^{(2)}(i\beta r), \qquad r > b. \tag{6.15}$$

To obtain the right mixture we examine the behavior of $h_l^{(1)}(\rho)$ and $h_l^{(2)}(\rho)$ for large values of ρ corresponding to the particle being located far from the center of force. In Chapter 10, we show that these functions have the asymptotic forms[4]

$$h_l^{(1)}(\rho) \xrightarrow[\rho \to \infty]{} -i \frac{e^{i(\rho - l\pi/2)}}{\rho}, \tag{6.16a}$$

and

$$h_l^{(2)}(\rho) \xrightarrow[\rho \to \infty]{} i \frac{e^{-i(\rho - l\pi/2)}}{\rho}. \tag{6.16b}$$

Since $\rho = i\beta r$, we see that as r goes to infinity $h_l^{(1)}(i\beta r)$ becomes vanishingly small, whereas $h_l^{(2)}(i\beta r)$ increases without limit. Therefore, in Eq. (6.15), we must choose $d_l' = 0$. For bound states $(E < 0)$ the complete radial wave function is

$$R_l(r) = \begin{cases} a_l j_l(\alpha r), & r \leq b, \\[2mm] c_l' h_l^{(1)}(i\beta r), & r > b. \end{cases} \tag{6.17}$$

---

[4]See Eqs. (10.26) and (10.27).

The energy spectrum follows from the requirement that the radial wave function $R_l(r)$ and its first derivative must be continuous[5] at the edge of the well where $r = b$.

$$a_l j_l(\alpha b) = c'_l h_l^{(1)}(i\beta b), \tag{6.18}$$

$$a_l \left[\frac{d}{dr} j_l(\alpha r)\right]_{r=b} = c'_l \left[\frac{d}{dr} h_l^{(1)}(i\beta r)\right]_{r=b}. \tag{6.19}$$

By taking the ratios of the two sides of these equations we eliminate the constants a_l and c'_l and obtain a transcendental equation for the energy E. We find that the energy is quantized for bound states. For a given l value, only certain values of E will satisfy this equation. The energy spectrum is not continuous; it is discrete.

The constants a_l and c'_l may be obtained from Eq. (6.18) and the requirement that $R_l(r)$ is normalized according to

$$\int_0^\infty [R_l(r)]^2 r^2 dr = 1.$$

6.1.2 CONTINUUM STATES $(E > 0)$

Here the corresponding classical particle has enough kinetic energy to escape from the well (see Figure 6.1c). For this case we take $\rho = \beta r$ with β defined in Eq. (6.13). Again, Eq. (6.5) follows with solutions

$$R_l(r) = c''_l h_l^{(1)}(\beta r) + d''_l h_l^{(2)}(\beta r), \qquad r > b. \tag{6.20}$$

From Eq. (6.16) we see that for large r this solution is a mixture of outgoing and incoming spherical waves. Since the continuity conditions at $r = b$ can be satisfied with *any* combination of $h_l^{(1)}$ and $h_l^{(2)}$ for the particle outside the potential well, the energy spectrum is continuous. It is *not* discrete for unbound states. However, for a *given* energy the boundary conditions will fix the mixture of incoming and outgoing waves for each l value.

6.2 Spherical Bessel Functions and Related Functions

Recursion formulas for the spherical Bessel functions may be obtained from the corresponding relations for the Bessel functions in Chapter 4. Direct substitution of Eqs. (6.9) into Eq. (4.21) yields

$$f_{l-1}(x) + f_{l+1}(x) = \frac{2l + 1}{x} f_l(x), \qquad l > 0, \tag{6.21}$$

---

[5]Continuity of the function and its first derivative is necessary since Eq. (6.1) is a *second-order* differential equation.

where $f_l(x)$ is either $j_l(x)$, $n_l(x)$, or any linear combination thereof. We differentiate Eqs. (6.9) and use Eq. (4.22) to obtain

$$f_l'(x) = \frac{l}{x} f_l(x) - f_{l+1}(x). \tag{6.22}$$

On combining this result with Eq. (6.21) we also have,

$$f_l'(x) = \frac{-(l+1)}{x} f_l(x) + f_{l-1}(x), \qquad l > 0. \tag{6.23}$$

Multiplying Eq. (6.22) by $l+1$ and Eq. (6.23) by l and then adding the two equations results in

$$(2l+1) f_l'(x) = l f_{l-1}(x) - (l+1) f_{l+1}(x), \qquad l > 0. \tag{6.24}$$

Now multiply Eq. (6.22) by x^{-l} and rearrange the terms. We find that

$$x^{-l} f_{l+1}(x) = l x^{-l-1} f_l(x) - x^{-l} f_l'(x),$$
$$= -\frac{d}{dx} \left[x^{-l} f_l(x) \right].$$

Thus,

$$\frac{d}{dx} \left[x^{-l} f_l(x) \right] = -x^{-l} f_{l+1}(x). \tag{6.25}$$

Similarly, from Eq. (6.23), we obtain

$$\frac{d}{dx} \left[x^{l+1} f_l(x) \right] = x^{l+1} f_{l-1}(x), \qquad l > 0. \tag{6.26}$$

Note that Eq. (6.25) may be rewritten as

$$\frac{f_{l+1}(x)}{x^{l+1}} = -\frac{1}{x} \frac{d}{dx} \left[\frac{f_l(x)}{x^l} \right].$$

Successive applications of this formula k times yields

$$\frac{f_l(x)}{x^l} = (-1)^k \left(\frac{1}{x} \frac{d}{dx} \right)^k \left[\frac{f_{l-k}(x)}{x^{l-k}} \right].$$

If $k = l$, we have

$$f_l(x) = (-1)^l x^l \left(\frac{1}{x} \frac{d}{dx} \right)^l [f_0(x)]. \tag{6.27}$$

This formula will make it possible for us to express the spherical Bessel functions and spherical Neumann functions of all orders in terms of trigonometric functions (see Eqs. 10.18 and 10.19).

6.3 Free Particle

An even simpler central force field is represented by a potential energy function

$$V(r) = 0 \quad \text{for all } r.$$

This corresponds to a *free* particle—a particle experiencing no force at all. On making the change of variable $\rho = kr$, Eq. (6.1) is transformed into Eq. (6.5) with the real parameter k defined by

$$k = (2mE/\hbar^2)^{1/2}.$$

Again the solution to Eq. (6.5) is given by Eq. (6.10). With the physical requirement that the solution be regular at the origin we have

$$R_l(r) = a_l j_l(kr), \quad 0 \le r \le \infty.$$

Thus, the free-particle eigenfunctions for definite orbital angular momentum l and projection m are

$$u_{lm}(\mathbf{r}) = a_l j_l(kr) Y_l^m(\theta, \phi). \tag{6.28}$$

These form a *complete set*[6] of eigensolutions of the free-particle Schrödinger equation.

6.4 Coulomb Field

An electron (electric charge $-e$) in the field of a point nucleus with charge Ze has potential energy[7]

$$V(r) = -\frac{1}{4\pi\varepsilon_0} \frac{Ze^2}{r}, \tag{6.29}$$

see Figure 6.2.

In this case the radial equation Eq. (6.1) becomes[8]

$$\frac{d^2}{dr^2}R + \frac{2}{r}\frac{d}{dr}R + \left[\frac{2mE}{\hbar^2} + \frac{2mZe^2}{4\pi\varepsilon_0\hbar^2 r} - \frac{l(l+1)}{r^2}\right]R = 0.$$

---

[6]See Chapter 12 for a discussion of *completeness*.

[7]In SI units, the force constant in Coulomb's law has the value $1/4\pi\varepsilon_0 = 9\times10^9$ newton-meter^2/coulomb^2.

[8]If we neglect the mass of the electron compared to that of the nucleus, then m is the electron mass.

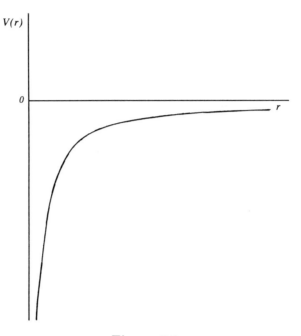

Figure 6.2

Again, we make the independent variable dimensionless by setting $\rho = \alpha r$ to obtain

$$\rho^2 \frac{d^2 R}{d\rho^2} + 2\rho \frac{dR}{d\rho} + \left[-\mu\rho^2 + \lambda\rho - l(l+1) \right] R = 0. \tag{6.30}$$

We have defined the parameters λ and μ by

$$\lambda = \frac{2mZe^2}{4\pi\varepsilon_0 \hbar^2 \alpha} \tag{6.31}$$

and

$$\mu = \frac{-2mE}{\hbar^2 \alpha^2}.$$

The real parameter α is yet to be defined. Note that $\lambda > 0$ and for bound states $(E < 0)$ we also have $\mu > 0$.

Because Eq. (6.30) has singularities at $\rho = 0$ and $\rho = \infty$, we expect to be able to transform it into the confluent hypergeometric equation. The zero-order term has three different powers of ρ. To eliminate two of these we look for a solution of the type

$$R(\rho) = u(\rho)v(\rho)e^{w(\rho)}. \tag{6.32}$$

On substituting this expression into Eq. (6.30) we obtain

$$
\begin{aligned}
\rho^2 v u'' &+ 2\rho[v + \rho w'v + \rho v']u' \\
&+ [\rho^2(w')^2 v + \rho^2 w''v + 2\rho^2 w'v' + \rho^2 v'' + 2\rho w'v + 2\rho v' \quad (6.33) \\
&- \mu\rho^2 v + \lambda\rho v - l(l+1)v]u = 0.
\end{aligned}
$$

To get rid of some of these terms we try $v = \rho^s$, in which case the zero-order term in Eq. (6.33) becomes

$$
\rho^s[\rho^2(w'' + (w')^2 - \mu) + \rho(2(s+1)w' + \lambda) + s(s+1) - l(l+1)]u. \quad (6.34)
$$

If we choose $s(s+1) = l(l+1)$, we see that $s = l$ or $s = -l - 1$. With $s = -l - 1$ it is clear from Eq. (6.32) that $R(\rho)$ diverges at $\rho = 0$, so we take $s = l$.

Now choose $w(\rho)$ so that the ρ^2 term in the square brackets in Eq. (6.34) drops out. This will give the zero-order term (as well as the second-order term) in Eq. (6.33) the desired form provided w' is independent of ρ. That is,

$$
w''(\rho) + (w'(\rho))^2 - \mu = 0,
$$

and

$$
w''(\rho) = 0.
$$

Hence, $w' = \pm\sqrt{\mu}$. The $+$ sign gives the wrong behavior of $R(\rho)$ for large ρ, so we take $w(\rho) = -\sqrt{\mu}\rho$, and, from Eq. (6.32),

$$
R(\rho) = \rho^l e^{-\sqrt{\mu}\rho} u(\rho). \quad (6.35)
$$

The differential equation for $u(\rho)$ (Eq. 6.33) now becomes

$$
\rho u''(\rho) + 2[(l+1) - \sqrt{\mu}\rho]u'(\rho) - [2(l+1)\sqrt{\mu} - \lambda]u(\rho) = 0. \quad (6.36)
$$

This is in the form of the confluent hypergeometric equation provided $2\sqrt{\mu} = 1$ in the first-order term. However, μ depends on the parameter α which is yet unspecified. Therefore, we choose α such that $\mu = \frac{1}{4}$, that is,

$$
\alpha = \sqrt{\frac{-8mE}{\hbar^2}}. \quad (6.37)
$$

With this choice Eq. (6.36) reduces to

$$
\rho u''(\rho) + [2(l+1) - \rho]u'(\rho) - [l + 1 - \lambda]u(\rho) = 0, \quad (6.38)
$$

which is the confluent hypergeometric equation. The general solution of Eq. (6.38) is

$$
u(\rho) = A\,_1F_1(l + 1 - \lambda; 2l + 2; \rho) + B\rho^{-(2l+1)}\,_1F_1(-l - \lambda; -2l; \rho).
$$

This function diverges at $\rho = 0$ unless $B = 0$. By following an argument similar to that leading to Eq. (3.10a), we find that

$$_1F_1(l+1-\lambda; 2l+2; \rho) \xrightarrow[\rho\to\infty]{} e^\rho.$$

Thus, $u(\rho)$ also diverges for large ρ unless $l + 1 - \lambda$ is a negative integer or zero, in which case the confluent hypergeometric function is a polynomial. This means that λ must be a positive integer, which we denote by n,

$$\lambda = n, \qquad n = 1, 2, 3, \ldots .$$

Thus, from Eq. (6.31) we find that the energy spectrum is discrete,

$$E_n = -\frac{Z^2 e^4 m}{32\pi^2 \varepsilon_0^2 \hbar^2 n^2}, \qquad n = 1, 2, 3, \ldots .$$

The corresponding radial solution (Eq. 6.35), which is regular at the origin and behaves properly for large values of ρ, is

$$R_{nl}(\rho) = A_{nl}\, \rho^l e^{-\frac{1}{2}\rho}\, {}_1F_1(-n+l+1; 2l+2; \rho), \tag{6.39}$$

where A_{nl} is a normalization constant. To obtain a polynomial solution we must have $n \geq l + 1$.

6.4.1 LAGUERRE POLYNOMIALS AND ASSOCIATED LAGUERRE FUNCTIONS

We note that Eq. (6.38) has the form

$$xu''(x) + (p+1-x)u'(x) + (q-p)u(x) = 0, \tag{6.40}$$

where p and q are integers. Since we are looking for polynomial solutions, we can define a new function $v(x)$ by

$$u(x) = \frac{d^p}{dx^p}\, v(x). \tag{6.41}$$

Clearly, if $u(x)$ is a polynomial, then $v(x)$ is also a polynomial.

On substituting this expression into Eq. (6.40) we find after some rearrangement,

$$\frac{d^p}{dx^p}\left[xv''(x) + (1-x)v'(x) + qv(x)\right] = 0.$$

We see that if we let $v(x)$ satisfy the differential equation

$$xv''(x) + (1-x)v'(x) + qv(x) = 0, \tag{6.42}$$

we get the solution to Eq. (6.40) given by Eq. (6.41). Note that Eq. (6.42) is also a confluent hypergeometric equation. This equation is known as *Laguerre's equation*. The solution that is regular at the origin is

$$v(x) = A_q \, {}_1F_1(-q; 1; x). \tag{6.43}$$

To have a polynomial solution q must be a positive integer. Then, in the series

$${}_1F_1(-q; 1; x) = \sum_{k=0}^{\infty} \frac{(-q)_k}{k!k!} x^k$$

all terms for $k > q$ vanish. Now,

$$\frac{(-q)_k}{k!} = (-1)^k \frac{(q - k + 1)_k}{k!}$$

$$= (-1)^k \sum_{m=0}^{k} \frac{(q + 1)_m}{m!} \frac{(-k)_{k-m}}{(k - m)!},$$

where we have invoked Vandermonde's theorem.

With this result we can write,

$${}_1F_1(-q; 1; x) = \sum_{k=0}^{\infty} \frac{(-1)^k}{k!} \left(\sum_{m=0}^{k} \frac{(q + 1)_m}{m!} \frac{(-k)_{k-m}}{(k - m)!} \right) x^k.$$

Now we use the identities

$$(m + q)! = q!(q + 1)_m$$

and

$$k! = (-1)^{k-m} m!(-k)_{k-m}, \qquad m \le k,$$

to obtain

$${}_1F_1(-q; 1; x) = \frac{1}{q!} \sum_{k=0}^{\infty} \left(\sum_{m=0}^{k} \frac{1}{(k - m)!} x^{k-m} \frac{(-1)^m (m + 1) \cdots (m + q)}{m!} x^m \right)$$

$$= \frac{1}{q!} \sum_{k=0}^{\infty} \left(\sum_{m=0}^{k} \frac{x^{k-m}}{(k - m)!} \frac{(-1)^m}{m!} \frac{d^q}{dx^q} x^{m+q} \right).$$

We recognize this last expression as the Cauchy product of two series. Thus,

$${}_1F_1(-q; 1; x) = \frac{1}{q!} \sum_{r=0}^{\infty} \frac{x^r}{r!} \frac{d^q}{dx^q} \sum_{s=0}^{\infty} \frac{(-1)^s}{s!} x^{s+q},$$

$$= \frac{1}{q!} e^x \frac{d^q}{dx^q} (x^q e^{-x}),$$

which is a polynomial of degree q.

We choose $A_q = q!$ in Eq. (6.43) to obtain for the polynomial solution of Laguerre's equation[9]

$$L_q(x) = q! \, _1F_1(-q; 1; x) = e^x \frac{d^q}{dx^q}(x^q e^{-x}). \tag{6.44}$$

This function is called the *Laguerre polynomial of order q*.

From Eq. (6.41), we see that the polynomial solution to Eq. (6.40) is obtained by differentiating $L_q(x)$ p times. We denote these solutions by

$$L_q^p(x) = \frac{d^p}{dx^p} L_q(x). \tag{6.45}$$

This function is known as the *associated Laguerre polynomial*. It is a polynomial of degree $q - p$.

By carrying out the differentiation explicitly, we find that

$$L_q^p(x) = q! \sum_{k=p}^{\infty} \frac{(-q)_k(k - p + 1)_p}{k!k!} x^{k-p}$$

$$= q! \sum_{k=0}^{\infty} \frac{(-q)_{k+p}(k + 1)_p}{(k + p)!(k + p)!} x^k,$$

where the second equality comes from replacing k by $k + p$. With the identities

$$(k + p)! = p!(p + 1)_k = k!(k + 1)_p$$

and

$$(-q)_{k+p} = (-q)_p(-q + p)_k,$$

we have

$$L_q^p(x) = \frac{q!(-q)_p}{p!} \sum_{k=0}^{\infty} \frac{(-q + p)_k}{k!(p + 1)_k} x^k. \tag{6.46}$$

This series is seen to be the confluent hypergeometric series. We generalize this result to include noninteger q and p and define the *associated Laguerre function* by,

$$L_q^p(x) = (-1)^p \frac{[\Gamma(q + 1)]^2}{\Gamma(p + 1)\Gamma(q - p + 1)} \, _1F_1(-q + p; p + 1; x). \tag{6.47}$$

---

[9]Some books on mathematical methods and special functions (e.g., N. Lebedev, *op. cit.*) define $L_q(x)$ consistent with $A_q = 1$. In all of the special functions in this book we use definitions that conform to standard texts on quantum mechanics (e.g., L.I. Schiff, *op. cit.*, and E. Merzbacher, *op. cit.*).

6.4.2 BOUND STATES IN A COULOMB FIELD

From Eq. (6.39) we see that the radial functions for bound states in a Coulomb potential are

$$R_{nl}(\rho) = N_{nl}\, \rho^l e^{-\frac{1}{2}\rho} L_{n+l}^{2l+1}(\rho), \qquad (6.48)$$

where N_{nl} is a normalization constant[10] and

$$\rho = \frac{Ze^2 mr}{2\pi\varepsilon_0 \hbar^2 n}. \qquad (6.49)$$

Continuum states for a particle in a Coulomb field are discussed in Chapter 10.

6.5 Isotropic Harmonic Oscillator

A third central force field, which often has applications in mathematical physics, is that of the spherically symmetric harmonic oscillator. The potential energy of a particle moving under the influence of such a force is

$$V(r) = \tfrac{1}{2} m\omega^2 r^2, \qquad (6.50)$$

where ω is a constant. The behavior of this field as a function of r is depicted in Figure 6.3.

On substituting this expression for $V(r)$ and changing to the dimensionless variable $\rho = r/b$, Eq. (6.1) becomes

$$\rho^2 \frac{d^2 R}{d\rho^2} + 2\rho \frac{dR}{d\rho} + \left[\lambda \rho^2 - \left(\frac{m\omega b^2}{\hbar}\right)^2 \rho^4 - l(l+1)\right] R = 0, \qquad (6.51)$$

where we have defined the parameter λ by

$$\lambda = \frac{2mEb^2}{\hbar^2}. \qquad (6.52)$$

The parameter b must have the dimension of a length, since ρ is dimensionless. If we choose

$$b = \sqrt{\hbar/m\omega}, \qquad (6.53)$$

---

[10] See Exercise 12.13.

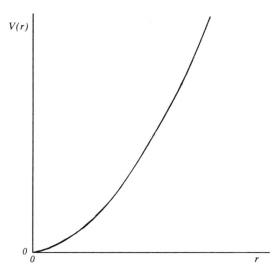

Figure 6.3

Eq. (6.51) then reads,

$$\rho^2 R''(\rho) + 2\rho R'(\rho) + \left[\lambda\rho^2 - \rho^4 - l(l+1)\right] R(\rho) = 0. \qquad (6.54)$$

From Eq. (6.52) we see that the energy spectrum is given by

$$E = \tfrac{1}{2}\lambda\hbar\omega, \qquad (6.55)$$

with the eigenvalues λ to be determined. The second-order differential equation for $R(\rho)$ has singular points at $\rho = 0$ and $\rho = \infty$, suggesting that we look for a change of variable, which transforms Eq. (6.54) into the confluent hypergeometric equation. Since the zero-order term has three powers of ρ (*viz.* 0, 2, and 4), we need to eliminate two of these. Toward this end we try

$$R(\rho) = u(\rho)v(\rho)e^{w(\rho)}. \qquad (6.56)$$

On substituting this expression into Eq. (6.54) we obtain a differential equation for $u(\rho)$,

$$\rho^2 v u''(\rho) + 2\rho\left[v + \rho w'v + \rho v'\right]u'(\rho) + \left[\rho^2(w')^2v + \rho^2 w''v\right.$$

$$\left. + 2\rho^2 w'v' + \rho^2 v'' + 2\rho w'v + 2\rho v' + \lambda\rho^2 v - \rho^4 v - l(l+1)v\right]u(\rho) = 0. \quad (6.57)$$

We wish to choose the functions $w(\rho)$ and $v(\rho)$ so that this reduces to the confluent hypergeometric equation. For one thing this means that the zero-order term in the equation for $u(\rho)$ must contain only one power of ρ. As a start we try $v(\rho) = \rho^s$, in which case the coefficient of $u(\rho)$ in Eq. (6.57) becomes

$$\rho^s \left[\rho^2 w'' + \rho^2 (w')^2 + 2(s+1)\rho w' + s(s+1) + \lambda \rho^2 - \rho^4 - l(l+1) \right].$$

We can eliminate the terms in ρ^{s+4} by taking $(w')^2 = \rho^2$, whence $w(\rho) = \pm \frac{1}{2}\rho^2$. In order for the radial wave function $R(\rho)$ to have the proper behavior for large ρ, we must take the minus sign, that is,

$$w(\rho) = -\tfrac{1}{2}\rho^2.$$

We now have for the coefficient of $u(\rho)$

$$\rho^s \left[(\lambda - 2s - 3)\rho^2 + s(s+1) - l(l+1) \right].$$

The terms in ρ^s can be eliminated by choosing $s(s+1) = l(l+1)$ or

$$s = \pm \left(l + \tfrac{1}{2} \right) - \tfrac{1}{2}.$$

The lower sign leads to solutions that are singular at the origin, so we take $s = l$. The radial function in Eq. (6.56) is now

$$R(\rho) = \rho^l e^{-\frac{1}{2}\rho^2} u(\rho) \tag{6.58}$$

and Eq. (6.57) reduces to

$$\rho u''(\rho) + 2(l + 1 - \rho^2)u'(\rho) - \rho(2l + 3 - \lambda)u(\rho) = 0. \tag{6.59}$$

On looking at the first-order term here, we see that the equation doesn't look like the confluent hypergeometric equation. The remedy consists in a change of the independent variable to a power of ρ, i.e., $x = \alpha \rho^n$. By following a procedure similar to that leading from Eq. (3.8) to Eq. (3.9) we find that $n = 2$ and $\alpha = 1$ yield the required result. Then Eq. (6.59) can be written as

$$x \frac{d^2}{dx^2} u(x) + \left(l + \tfrac{3}{2} - x \right) \frac{d}{dx} u(x) - \tfrac{1}{4}(2l + 3 - \lambda)u(x) = 0,$$

which is the confluent hypergeometric equation. The general solution is

$$u(x) = A \,_1F_1 \left(\tfrac{1}{2} \left(l + \tfrac{3}{2} \right) - \tfrac{\lambda}{4}; l + \tfrac{3}{2}; x \right)$$
$$+ Bx^{-l-\frac{1}{2}} \,_1F_1 \left(-\tfrac{1}{2} \left(l - \tfrac{1}{2} \right) - \tfrac{\lambda}{4}; -l + \tfrac{1}{2}; x \right).$$

Since the second term is singular at $x = 0$ ($\rho = 0$), we must set $B = 0$. From an argument like the one leading to Eqs. (3.10) we conclude that $R(\rho)$ in Eq. (6.56) diverges for large ρ unless the confluent hypergeometric series is a polynomial, that is

$$u(\rho) = A \, _1F_1 \left(-n; l + \tfrac{3}{2}; \rho^2 \right), \tag{6.60}$$

where

$$n = \tfrac{1}{4}(\lambda - 2l - 3) \tag{6.61}$$

is a nonnegative integer.

On comparing Eq. (6.60) with Eq. (6.47) we recognize that $u(\rho)$ may also be written as an associated Laguerre polynomial, and the radial wave function in Eq. (6.54) becomes

$$R_{nl}(\rho) = N_{nl}\rho^l e^{-\frac{1}{2}\rho^2} L_{n+l+\frac{1}{2}}^{l+\frac{1}{2}} (\rho^2). \tag{6.62}$$

Now, we can solve Eq. (6.61) for λ and find that the energy eigenvalues in Eq. (6.55) are

$$E_n = \left(2n + l + \tfrac{3}{2} \right) \hbar\omega. \tag{6.63}$$

Calculation of the normalization constant N_{nl} is left as an exercise.[11]

6.6 Special Functions and Hypergeometric Functions

In the last four chapters, we have considered the special mathematical functions that are encountered most frequently by engineers and physicists. We have shown these functions arising in a variety of physical contexts and have defined all of them in terms of the generalized hypergeometric function. These results are summarized in Table 6.1. In subsequent chapters we show that these are equivalent to other definitions commonly found in works on physics and applied mathematics. We shall also examine some of the important properties of these functions.

---

[11]See Exercise 12.11.

TABLE 6.1. Special functions and hypergeometric functions

Special Function	Symbol	Definition
Hermite	$H_n(z)$	$\dfrac{n!(-1)^{-\frac{1}{2}n}}{(n/2)!}\, {}_1F_1\left(-\frac{n}{2};\frac{1}{2};z^2\right)$ (even n) $\dfrac{2n!(-1)^{\frac{1}{2}(1-n)}}{[(n-1)/2]!}\, z\, {}_1F_1\left(-\frac{n-1}{2};\frac{3}{2};z^2\right)$ (odd n)
Bessel	$J_\nu(z)$	$\dfrac{e^{-iz}}{\Gamma(\nu+1)}\left(\dfrac{z}{2}\right)^\nu {}_1F_1\left(\nu+\frac{1}{2};2\nu+1;2iz\right)$
Legendre	$P_n(z)$	${}_2F_1\left(-n,n+1;1;\frac{1}{2}(1-z)\right)$
Associated Legendre	$P_n^m(z)$	$\dfrac{(n+|m|)!(1-z^2)^{\frac{1}{2}|m|}}{2^{|m|}|m|!(n-|m|)!}$ $\times\, {}_2F_1\left(-n+|m|,n+|m|+1;|m|+1;\frac{1}{2}(1-z)\right)$
Chebyshev Type 1	$T_n^{(1)}(z)$	${}_2F_1\left(-n,n;\frac{1}{2};\frac{1}{2}(1-z)\right)$
Chebyshev Type 2	$T_n^{(2)}(z)$	$(n+1)\, {}_2F_1\left(-n,n+2;\frac{3}{2};\frac{1}{2}(1-z)\right)$
Laguerre	$L_n(z)$	$n!\, {}_1F_1(-n;1;z)$
Associated Laguerre	$L_n^p(z)$	$(-1)^p\dfrac{[\Gamma(n+1)]^2}{\Gamma(p+1)\Gamma(n-p+1)}\, {}_1F_1(-n+p;p+1;z)$

EXERCISES

1. Show that for a particle in a central force field the radial wave equation for states with zero angular momentum ($l = 0$) can be written as

$$\frac{d^2}{dr^2}g(r) + \frac{2m}{\hbar^2}[E - V(r)]g(r) = 0,$$

   where $g(r)$ is related to the radial function in Eq. (6.1) by

$$g(r) = rR(r).$$

2. Show that consistent with Eq. (6.9b) the spherical Neumann function is also defined by

$$n_l(\rho) = \sqrt{\frac{\pi}{2\rho}} N_{l+\frac{1}{2}}(\rho).$$

3. Consider a particle of mass m bound in a central force field with the potential energy of the particle given by

$$V(r) = -V_0 e^{-r/b},$$

   where b is a constant length. Make a change of independent variable to

$$\rho = \lambda e^{-\mu r}$$

   with parameters λ and μ to be determined. Find the values of λ and μ such that the radial equation for bound states ($E < 0$) with zero angular momentum (see Exercise 6.1) can be written as

$$\rho^2 \frac{d^2}{d\rho^2}g(\rho) + \rho \frac{d}{d\rho}g(\rho) + (\rho^2 - \nu^2)u(\rho) = 0.$$

   This is Bessel's equation, which has the general solution

$$g(\rho) = AJ_\nu(\rho) + BN_\nu(\rho).$$

   Invoke the boundary conditions on the radial wave function $R(r)$ at $r = \infty$ to show that $B = 0$. Show also that the spectrum of energy eigenstates with $l = 0$ is determined by the condition

$$J_\nu\left(\sqrt{\frac{8mV_0 b^2}{\hbar^2}}\right) = 0.$$

   That is, given the well parameters b and V_0 Bessel functions $J_\nu(\rho)$ of all orders ν that have zeros at $\rho = \sqrt{8mV_0 b^2/\hbar^2}$ correspond to zero angular momentum energy eigenstates with radial wave functions

$$R(r) = \frac{A}{r}J_\nu(\rho), \qquad l = 0.$$

   Thus, the energy spectrum for these states is discrete. Write down the energy in terms of the parameter ν.

7

Complex Analysis

As we have already seen, complex numbers crop up quite naturally in physics. Often, the solutions to problems are conveniently expressed in terms of complex functions [e.g., the angular momentum eigenfunctions $Y_l^m(\theta, \phi)$ of Eq. (5.26)]. Here, we employ the techniques of complex analysis to establish the equivalence of various definitions of each special function. Most physics and applied mathematics texts in which special functions appear make use of generating functions, asymptotic forms, and recursion formulas for these functions. The properties of analytic functions will be useful to us in deriving these relations. In the next two chapters we develop enough of the theory of complex variables to enable us to carry out these tasks.

7.1 Complex Numbers

To begin our discussion of complex analysis let us review some essential features of complex arithmetic. Consider the equation

$$x^2 + 1 = 0. \tag{7.1}$$

Within the realm of real numbers there is no solution to this equation, since there is no real number x such that $x^2 = -1$. In order to obtain solutions to such equations we introduce a number i whose square is -1,

$$i = \sqrt{-1}.$$

Thus, the two solutions to the quadratic equation (Eq. 7.1) are $x = \pm i$.
Similarly, the equation
$$x^2 + 5 = 0$$
has solutions
$$x = \pm i\sqrt{5}.$$

Such numbers, which lie outside the domain of real numbers, are said to be *imaginary*. The general form of an imaginary number λ is

$$\lambda = ib,$$

where b is real.

Now consider the equation

$$x^2 - 2x + 2 = 0, \tag{7.2}$$

which, when rewritten as

$$(x - 1)^2 = -1,$$

is seen to have the solutions $x - 1 = \pm i$ or

$$x = 1 \pm i,$$

the sum of a real number and an imaginary number. Such numbers are said to be *complex*. The general form of a complex number z is

$$z = x + iy,$$

where x and y are real with x being the *real part* of z and y the *imaginary part* of z.[1] Corresponding to each complex number z, there is a complex number $z^* = x - iy$ called the *complex conjugate* of z.

Clearly, any complex number z can be expressed as an ordered pair of real numbers (x, y). Such a number can be represented graphically as a

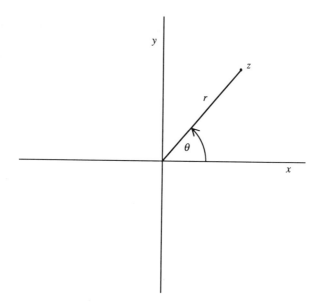

Figure 7.1

---

[1]We shall also denote the real and imaginary parts of z by $\mathrm{Re}(z)$ and $\mathrm{Im}(z)$, respectively.

point in a plane. In this *complex plane* we often find it convenient to represent complex numbers in terms of polar coordinates (r, θ) as well as the rectangular ones (x, y); see Figure 7.1. These sets of coordinates are related by

$$x = r \cos \theta \quad \text{and} \quad y = r \sin \theta,$$

or by

$$r = \sqrt{x^2 + y^2} \quad \text{and} \quad \theta = \tan^{-1}(y/x).$$

On writing the complex number z in polar coordinates we have

$$z = r(\cos \theta + i \sin \theta). \tag{7.3}$$

The *absolute value* of z is defined by

$$|z| = |x + iy| = \sqrt{x^2 + y^2}.$$

Clearly, $|z| = r$. The polar angle θ is called the *argument* of z and is often denoted by $\arg z$. In measuring θ the counterclockwise direction is taken as positive.

7.2   Analytic Functions of a Complex Variable

Let us now generalize the definition of an *analytic function* to complex variables.

Definition. Let $f(z)$ be a function of the complex variable z. If the derivative of $f(z)$ exists everywhere in a small neighborhood around a point z_0 including z_0 itself, then $f(z)$ is *analytic* at z_0. If $f'(z)$ does not exist at $z = z_0$, then z_0 is a singular point.

This definition of analyticity depends on the definition of the derivative of a complex function. Any complex function $f(z)$ may be written expressly in terms of its real and imaginary parts,

$$f(z) = u(x, y) + iv(x, y), \tag{7.4}$$

where u and v are real functions of the real variables x and y.

Example. Suppose $f(z) = z^2$. Then

$$f(z) = (x + iy)^2,$$
$$= x^2 - y^2 + i2xy.$$

The real and imaginary parts of $f(z)$ are, respectively

$$u(x, y) = x^2 - y^2 \quad \text{and} \quad v(x, y) = 2xy.$$

7.2.1 THE CAUCHY–RIEMANN EQUATIONS

We define the derivative of $f(z)$ as

$$f'(z) = \frac{df}{dz} = \lim_{\Delta z \to 0} \frac{f(z + \Delta z) - f(z)}{(z + \Delta z) - z} = \lim_{\Delta z \to 0} \frac{\Delta f}{\Delta z}. \qquad (7.5)$$

The increments Δz and Δf have the meaning

$$\Delta z = \Delta x + i\Delta y \quad \text{and} \quad \Delta f = \Delta u + i\Delta v$$

so that

$$\frac{\Delta f}{\Delta z} = \frac{\Delta u + i\Delta v}{\Delta x + i\Delta y}.$$

The limit in Eq. (7.5) must be independent of the approach to the point z. Let us take two *different* approaches (which are not, of course, the only ones possible):

1. Set $\Delta y = 0$ and take the limit as $\Delta x \to 0$ (hence $\Delta z = \Delta x$)

$$\lim_{\Delta z \to 0} \left[\frac{\Delta f}{\Delta z} \right]_{\Delta y = 0} = \lim_{\Delta x \to 0} \left[\frac{\Delta u}{\Delta x} + \frac{i\Delta v}{\Delta x} \right]$$

$$= \frac{\partial u}{\partial x} + i\frac{\partial v}{\partial x}.$$

2. Set $\Delta x = 0$ and take the limit as $\Delta y \to 0$ (hence $\Delta z = i\Delta y$)

$$\lim_{\Delta z \to 0} \left[\frac{\Delta f}{\Delta z} \right]_{\Delta x = 0} = \lim_{\Delta y \to 0} \left[\frac{\Delta u}{i\Delta y} + \frac{i\Delta v}{i\Delta y} \right]$$

$$= -i\frac{\partial u}{\partial y} + \frac{\partial v}{\partial y}.$$

Since the derivative $f'(z)$ is unique, that is, independent of the approach to the limit, then the real parts in the two approaches above must be equal. Similarly, the imaginary parts must also be equal. Therefore,

$$\frac{\partial u}{\partial x} = \frac{\partial v}{\partial y} \quad \text{and} \quad \frac{\partial v}{\partial x} = -\frac{\partial u}{\partial y}. \qquad (7.6)$$

These are called the *Cauchy–Riemann* equations. They represent a *necessary* condition for the existence of the derivative $f'(z)$. That is, if $f'(z)$ exists, then the Cauchy–Riemann relations hold.

We now show that the Cauchy–Riemann equations together with the continuity of the partial derivatives $\partial f / \partial x$ and $\partial f / \partial y$ provide us with a *sufficient* condition for the existence of $f'(z)$. We see from Eq. (7.4) that $f(z)$ is a function of the two variables x and y. Thus,

$$\Delta f = \frac{\partial f}{\partial x} \Delta x + \frac{\partial f}{\partial y} \Delta y$$

$$= \frac{\partial u}{\partial x} \Delta x + i\frac{\partial v}{\partial x} \Delta x + \frac{\partial u}{\partial y} \Delta y + i\frac{\partial v}{\partial y} \Delta y.$$

Now divide by $\Delta z = \Delta x + i \Delta y$ to obtain

$$\frac{\Delta f}{\Delta z} = \left[\frac{\partial u}{\partial x} + i \frac{\partial v}{\partial x} \right] \frac{\Delta x}{\Delta x + i \Delta y} + \left[\frac{\partial u}{\partial y} + i \frac{\partial v}{\partial y} \right] \frac{\Delta y}{\Delta x + i \Delta y}$$

$$= \frac{1}{1 + i \Delta y / \Delta x} \left[\frac{\partial u}{\partial x} + i \frac{\partial v}{\partial x} + \frac{\Delta y}{\Delta x} \left(\frac{\partial u}{\partial y} + i \frac{\partial v}{\partial y} \right) \right].$$

By using the Cauchy–Riemann equations for the two terms in parentheses in this last expression, we can factor out of the square brackets the factor $1 + i \Delta y / \Delta x$, which cancels the same factor in the denominator. This gives

$$\frac{\Delta f}{\Delta z} = \frac{\partial u}{\partial x} + i \frac{\partial v}{\partial x}$$

$$= \frac{\partial}{\partial x} (u + iv) = \frac{\partial f}{\partial x}.$$

In passing to the limit we see that

$$\lim_{\Delta z \to 0} \frac{\Delta f}{\Delta z} = \frac{\partial f}{\partial x}$$

is *independent* of the approach to the limit if the Cauchy–Riemann relations hold and $\partial f / \partial x$ is continuous.

We also find from the Cauchy–Riemann equations that

$$\frac{\partial f}{\partial x} = \frac{\partial u}{\partial x} + i \frac{\partial v}{\partial x} = \frac{\partial v}{\partial y} + i \left(-\frac{\partial u}{\partial y} \right) = -i \frac{\partial f}{\partial y}.$$

Thus, if the Cauchy–Riemann equations hold and $\partial f / \partial x$ is continuous, then $\partial f / \partial y$ is also continuous. We now have the sufficient condition we sought. If the Cauchy–Riemann equations hold and the partial derivatives $\partial f / \partial x$ and $\partial f / \partial y$ are continuous, then $f'(z)$ exists and $f(z)$ is analytic.

7.2.2 The Cauchy Integral Theorem

We can obtain a useful integral theorem for a function of a complex variable by making use of the following theorem. For two functions $P(x, y)$ and $Q(x, y)$ which have continuous first derivatives we have

$$\oint_C (P dx + Q dy) = \int_S \left(\frac{\partial Q}{\partial x} - \frac{\partial P}{\partial y} \right) dx dy, \tag{7.7}$$

where S is the region in the x-y plane bounded by the closed curve C. This is *Green's theorem*.

Our proof of Green's theorem is based on the curve in Figure 7.2a. The result is easily extended to curves of more general shape by recognizing that distortions can be treated by decomposing the region into subregions like the one shown. Let $x_1(y)$ and $x_2(y)$ be the x coordinates of the two

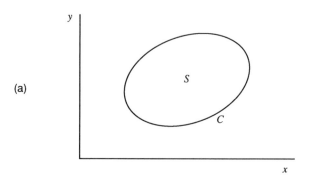

(a)

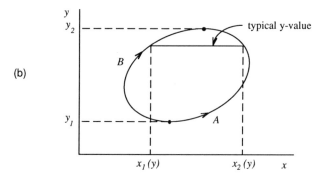

(b)

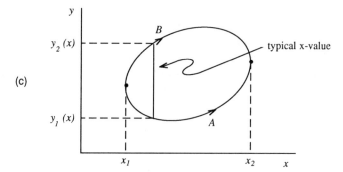

(c)

Figure 7.2

points on the curve for a given value of y, as in Figure 7.2b. Clearly, integration of $\partial Q/\partial x$ over S gives

$$
\begin{aligned}
\int_S \frac{\partial Q}{\partial x} \, dx \, dy &= \int_{y_1}^{y_2} \left[\int_{x_1(y)}^{x_2(y)} \frac{\partial Q}{\partial x} \, dx \right] dy \\
&= \int_{y_1}^{y_2} \left[Q(x_2(y), y) - Q(x_1(y), y) \right] dy \\
&= \int_{y_1 \text{ (path } A)}^{y_2} Q \, dy - \int_{y_1 \text{ (path } B)}^{y_2} Q \, dy.
\end{aligned}
$$

Now, let us reverse the direction of integration along path B and make a corresponding sign change in the second integral to get

$$
\int_S \frac{\partial Q}{\partial x} \, dx \, dy = \oint_C Q(x, y) \, dy. \tag{7.8}
$$

In a similar fashion we see from Figure 7.2c that

$$
\begin{aligned}
\int_S \frac{\partial P}{\partial y} \, dx \, dy &= \int_{x_1}^{x_2} \left[\int_{y_1(x)}^{y_2(x)} \frac{\partial P}{\partial y} \, dy \right] dx \\
&= \int_{x_1 \text{ (path } A)}^{x_2} P(x, y_2(x)) \, dx - \int_{x_1 \text{ (path } B)}^{x_2} P(x, y_1(x)) \, dx.
\end{aligned}
$$

By integrating from x_2 to x_1 on path A and changing the sign of the corresponding integral we have

$$
\int_S \frac{\partial P}{\partial y} \, dx \, dy = - \oint_C P(x, y) \, dx. \tag{7.9}
$$

Now subtract Eq. (7.9) from Eq. (7.8) and obtain

$$
\oint_C (P \, dx + Q \, dy) = \int_S \left[\frac{\partial Q}{\partial x} - \frac{\partial P}{\partial y} \right] dx \, dy, \tag{7.10}
$$

which is Green's theorem.

Next, we introduce the notion of a *simply connected* region in the complex plane. We say that a region is simply connected if, for every closed curve in the region, the area bounded by the curve, including the curve itself, lies wholly within the region. In other words, the region has no holes in it; see Figure 7.3.

Let R be a simply connected region in the complex plane. If $f(z)$ is analytic in R, then

$$
\oint_C f(z) \, dz = 0, \tag{7.11}
$$

simply-connected

doubly-connected

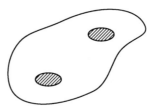

triply-connected

Figure 7.3

for every closed curve C in R. This is the *Cauchy integral theorem*. To prove it we write

$$\oint_C f(z)\,dz = \oint_C (u + iv)(dx + idy)$$

$$= \oint_C (u\,dx - v\,dy) + i \oint_C (v\,dx + u\,dy)$$

$$= \int_S \left(-\frac{\partial v}{\partial x} - \frac{\partial u}{\partial y}\right) dxdy + i \int_S \left(\frac{\partial u}{\partial x} - \frac{\partial v}{\partial y}\right) dxdy,$$

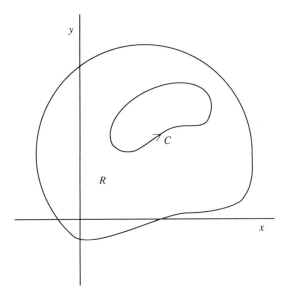

Figure 7.4

where we have invoked Green's theorem assuming that there are no holes in R. See Figure 7.4. If we use the Cauchy–Riemann equations in this last expression, we find that the real and imaginary parts vanish separately and the theorem is proved.[2]

By means of a *cut line*, we can convert a multiply connected region into a simply connected one. In Figure 7.5a the two curves C_1 and C_2 enclose the same hole. However, the region *bounded* by these curves has *no* holes, hence it is simply connected.

We form a single closed curve C from C_1 and C_2 as shown in Figure 7.5b. That is,

$$C = C_1 + (-C_2) + \text{paths parallel to cut line.}$$

By allowing the separation between the parallel paths on each side of the cut to become vanishingly small the contributions from these two pieces cancel each other. Thus,

---

[2]Green's theorem requires that the derivatives of u and v be continuous on and inside C. This is true for the functions we consider in this book. For a proof of Cauchy's integral theorem, which does not depend on this restriction see R.V. Churchill, *Introduction to Complex Variables and Applications*, McGraw-Hill, New York, 1948, p. 83.

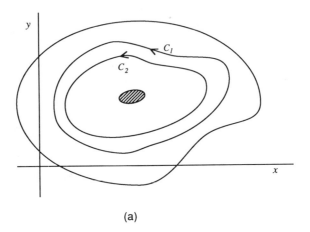

(a)

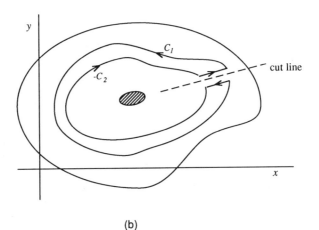

(b)

Figure 7.5

$$\oint_C f(z)\,dz = \oint_{C_1} f(z)\,dz + \oint_{-C_2} f(z)\,dz$$

$$= \oint_{C_1} f(z)\,dz - \oint_{C_2} f(z)\,dz,$$

since the curves close as the cut gap vanishes. On invoking the Cauchy integral theorem we see that

$$\oint_{C_1} f(z)\,dz = \oint_{C_2} f(z)\,dz.$$

Hence, the value of the integral is the same for *all* closed paths surrounding the hole. The value is independent of the contour.

7.2.3 THE CAUCHY INTEGRAL FORMULA

A function may be analytic at all points in some neighborhood of a point z_0, but not at z_0 itself. In this case the function is said to have an *isolated singularity* at z_0.

Suppose the function $f(z)$ is analytic on the closed contour C and everywhere inside the region bounded by C. Then, the function

$$\frac{f(z)}{z - z_0}$$

is analytic everywhere in this region except for an isolated singularity at z_0 as shown in Figure 7.6a. For such a function $f(z)$ we have the very useful formula

$$\oint_C \frac{f(z)}{z - z_0} \, dz = 2\pi i f(z_0). \tag{7.12}$$

This is the *Cauchy integral formula*. To derive it we construct a contour as shown in Figure 7.6b. The curve C' is a small circle of radius r which surrounds z_0 in a clockwise direction. Since the region bounded by the complete contour is simply connected, we have from the Cauchy integral theorem,

$$\oint_{C+C'+\text{cut}} \frac{f(z)}{z - z_0} \, dz = 0.$$

Again the contributions along the cut cancel as the cut gap becomes vanishingly small and we have

$$\oint_C \frac{f(z)}{z - z_0} \, dz = - \oint_{C'} \frac{f(z)}{z - z_0} \, dz.$$

On C' $z = z_0 + r(\cos\theta + i\sin\theta)$ which leads to

$$\oint_{C'} \frac{f(z)}{z - z_0} \, dz = i \oint_{C'} f\big(z_0 + r(\cos\theta + i\sin\theta)\big) \, d\theta.$$

Now we take the limit as r goes to 0 and see that this equation reduces to

$$\oint_{C'} \frac{f(z)}{z - z_0} \, dz = -2\pi i f(z_0).$$

The minus sign arises from the clockwise integration. Thus, the result in Eq. (7.12) is proved.

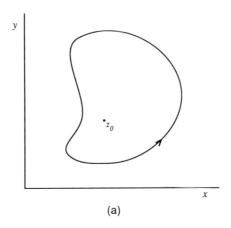

(a)

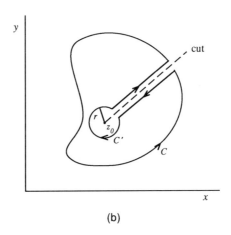

(b)

Figure 7.6

7.3 Analyticity

According to our definition in Chapter 2, a function that is analytic in some interval may be represented by a power series expanded about any point in the interval. In this chapter the definition of analyticity is stated somewhat differently, *viz.* a function $f(z)$ is analytic at a point $z = z_0$, if its derivative exists at z_0 and everywhere in a small neighborhood around z_0. To demonstrate that these two definitions are consistent, we show that any function that can be represented by a power series within its circle of convergence has a derivative and is therefore analytic in the sense we have introduced here.

First, we show that within the circle of convergence a power series converges uniformly. Suppose

$$S(z) = \sum_{k=0}^{\infty} a_k z^k$$

converges for $|z| < R$. Let R_1 and r be real positive numbers such that

$$R_1 < r < R.$$

Clearly, the series converges for $|z| \leq R_1$ and for $z = r$. Because $S(r)$ converges, a finite number K exists independent of k such that

$$|a_k r^k| < K.$$

By constructing the nth partial sum $S_n(z)$ we find that

$$|S(z) - S_n(z)| = \left| \sum_{k=n+1}^{\infty} a_k z^k \right| \leq \sum_{k=n+1}^{\infty} |a_k z^k|$$

$$< \sum_{k=n+1}^{\infty} K \left(\frac{R_1}{r} \right)^k = \frac{K \left(\frac{R_1}{r} \right)^{n+1}}{1 - \frac{R_1}{r}} = \varepsilon(n).$$

Thus, given any ε we can find a corresponding value of n independent of z for any z with $|z| \leq R_1$ such that the inequality above holds. Hence, the series $S(z)$ converges uniformly.

Now let us show that a function represented by a convergent power series is analytic, i.e., has a derivative. Suppose

$$f(z) = \sum_{n=0}^{\infty} a_n z^n$$

converges for $|z| < R$. Since this is a power series, it converges uniformly and we can differentiate it term by term to obtain

$$g(z) = \sum_{n=1}^{\infty} n a_n z^{n-1}.$$

If $|z + \Delta z| < R$, then

$$\frac{f(z + \Delta z) - f(z)}{\Delta z} - g(z) = \sum_{n=0}^{\infty} a_n \left[\frac{(z + \Delta z)^n - z^n}{\Delta z} - n z^{n-1} \right].$$

Furthermore,

$$\left| \frac{f(z + \Delta z) - f(z)}{\Delta z} - g(z) \right| \leq \sum_{n=0}^{\infty} |a_n| \left(\frac{(z + \Delta z)^n - z^n}{\Delta z} - n z^{n-1} \right) \right|.$$

By using the binomial expansion for $(z + \Delta z)^n$ we get

$$\left| \frac{(z + \Delta z)^n - z^n}{\Delta z} - nz^{n-1} \right| = \left| \sum_{k=2}^{n} \frac{n!}{k!(n-k)!} z^{n-k} \Delta z^{k-1} \right|$$

$$\leq \sum_{k=2}^{n} \frac{n!}{k!(n-k)!} |z|^{n-k} |\Delta z|^{k-1}$$

$$= \frac{1}{|\Delta z|} \left[(|z| + |\Delta z|)^n - |z|^n \right] - n|z|^{n-1}.$$

Thus,

$$\left| \frac{f(z + \Delta z) - f(z)}{\Delta z} - g(z) \right|$$

$$\leq \sum_{n=0}^{\infty} \left| a_n \left(\frac{(|z| + |\Delta z|)^n - |z|^n}{|\Delta z|} - n|z|^{n-1} \right) \right|.$$

Let r be a real, positive number such that $|z| + |\Delta z| < r < R$. Then,

$$\left| \frac{f(z + \Delta z) - f(z)}{\Delta z} - g(z) \right|$$

$$\leq \sum_{n=0}^{\infty} \left| a_n r^n \left(\frac{1}{r^n} \left\{ \frac{(|z| + |\Delta z|)^n - |z|^n}{|\Delta z|} - n|z|^{n-1} \right\} \right) \right|.$$

The series representing $f(z)$ converges uniformly. Therefore, there exists some number N independent of n such that $|a_n r^n| \leq N$. Hence,

$$\left| \frac{f(z + \Delta z) - f(z)}{\Delta z} - g(z) \right|$$

$$\leq N \sum_{n=0}^{\infty} \left[\frac{1}{|\Delta z|} \left(\frac{|z| + |\Delta z|}{r} \right)^n - \frac{1}{|\Delta z|} \left(\frac{|z|}{r} \right)^n - \frac{n|z|^{n-1}}{r^n} \right].$$

We evaluate the sums in this last expression and find that

$$\left| \frac{f(z + \Delta z) - f(z)}{\Delta z} - g(z) \right| \leq \frac{Nr|\Delta z|}{(r - |z| - |\Delta z|)(r - |z|)^2},$$

where the right-hand side clearly goes to zero as $\Delta z \to 0$. Thus, the derivative of $f(z)$ is

$$\lim_{\Delta z \to 0} \left[\frac{f(z + \Delta z) - f(z)}{\Delta z} \right] = g(z).$$

The derivative of $f(z)$ exists for $|z| \leq r < R$, therefore, $f(z)$ is analytic within its circle of convergence $|z| < R$ as we wished to show.

Next, we show that the converse is also true, namely that an analytic function $f(z)$ may be represented by its Taylor series within its circle of convergence. First, we use Eq. (7.12) to express the derivatives of $f(z)$ in terms of contour integrals,

$$f'(z) = \lim_{\Delta z \to 0} \frac{f(z + \Delta z) - f(z)}{\Delta z}$$

$$= \lim_{\Delta z \to 0} \frac{1}{\Delta z} \left[\frac{1}{2\pi i} \oint_C \frac{f(t)}{t - z - \Delta z} \, dt - \frac{1}{2\pi i} \oint_C \frac{f(t)}{t - z} \, dt \right]$$

$$= \lim_{\Delta z \to 0} \frac{1}{2\pi i} \oint_C \frac{f(t)}{(t - z)(t - z - \Delta z)} \, dt.$$

Thus,

$$f'(z) = \frac{1}{2\pi i} \oint_C \frac{f(t)}{(t - z)^2} \, dt. \tag{7.13}$$

Similarly, for the second derivative we have

$$f''(z) = \lim_{\Delta z \to 0} \frac{f'(z + \Delta z) - f'(z)}{\Delta z}$$

$$= \lim_{\Delta z \to 0} \frac{1}{2\pi i \Delta z} \left[\oint_C \frac{f(t)}{(t - z - \Delta z)^2} \, dt - \oint_C \frac{f(t)}{(t - z)^2} \, dt \right]$$

$$= \frac{2}{2\pi i} \oint_C \frac{f(t)}{(t - z)^3} \, dt.$$

In arriving at this result we have used Eq. (7.13).

We continue this procedure to obtain for the nth derivative of $f(z)$

$$f^{(n)}(z) = \frac{n!}{2\pi i} \oint_C \frac{f(t)}{(t - z)^{n+1}} \, dt. \tag{7.14}$$

Suppose $f(t)$ is single-valued and has a derivative everywhere in a given region R. Let C be a circle of radius $r = |t - z_0|$ centered on the point z_0, as in Figure 7.7. Consider a point z inside the circle. Then, by the Cauchy integral formula

$$f(z) = \frac{1}{2\pi i} \oint_C \frac{f(t)}{t - z} \, dt, \tag{7.15}$$

with $|z - z_0| < r$.

The denominator in the integrand may be rewritten as

$$\frac{1}{t - z} = \frac{1}{t - z_0} \left[\frac{1}{1 - x} \right],$$

with $x = (z - z_0)/(t - z_0)$. Because $|x| < 1$, we can write this result in terms of the geometric series,[3]

$$\frac{1}{t - z} = \sum_{k=0}^{\infty} \frac{(z - z_0)^k}{(t - z_0)^{k+1}}.$$

---

[3] See Eq. (2.6).

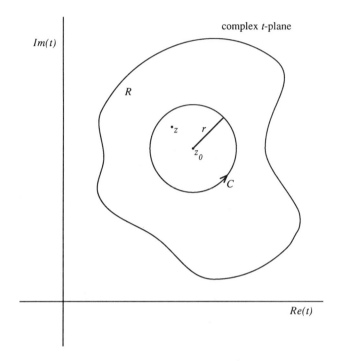

Figure 7.7

On inserting this result into Eq. (7.15) and interchanging the order of summation and integration we get

$$f(z) = \frac{1}{2\pi i} \sum_{k=0}^{\infty} \left[\left(\oint_C \frac{f(t)}{(t - z_0)^{k+1}} \, dt \right) (z - z_0)^k \right],$$

which, when combined with Eq. (7.14), gives

$$f(z) = \sum_{k=0}^{\infty} \frac{f^{(k)}(z_0)}{k!} (z - z_0)^k. \qquad (7.16)$$

This is exactly the form of Eq. (2.2) defining the Taylor series.

The Taylor formula with remainder also holds for a function of a complex variable. The generalization of Eq. (2.12) to a function of a complex variable z expanded about a point z_0 in the complex plane is

$$f(z) = \sum_{k=0}^{n-1} \frac{f^{(k)}(z_0)}{k!} (z - z_0)^k + \frac{(z - z_0)^n}{2\pi i} \oint_C \frac{f(t)}{(t - z_1)^{n+1}} \, dt, \qquad (7.17)$$

which we give without proof. The point z_1 satisfies the condition $|z_1 - z_0| < |z - z_0|$ and the contour C is shown in Figure 7.8.

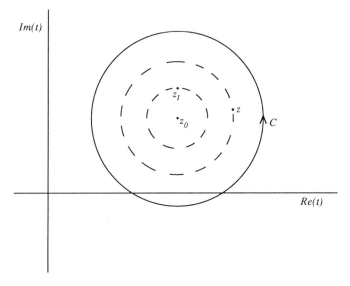

Figure 7.8

7.3.1   ELEMENTARY FUNCTIONS

It is clear that the polynomial

$$f(z) = \sum_{k=0}^{n} a_k z^k$$

is analytic for all z. Also the rational function

$$f(z) = \sum_{k=0}^{n} a_k z^k \bigg/ \sum_{j=0}^{m} b_j z^j$$

is analytic in any region that does not contain a zero of the denominator.

Now, let us extend the definition of the exponential function to a function of a complex variable z. We define

$$e^z = e^x (\cos y + i \sin y), \tag{7.18}$$

which is analytic for all z.[4] Note that for $y = 0$ this function reduces to the real exponential e^x, which we already know. If $x = 0$ we have Euler's formula,

$$e^{iy} = \cos y + i \sin y.$$

---

[4]See Exercise 7.10.

This latter expression allows us to write the polar representation of z (Eq. 7.3) in a more compact form,

$$z = re^{i\theta}.$$

By analogy with real trigonometric and hyperbolic functions we make the following definitions:

$$\sin z = \tfrac{1}{2i}\left[e^{iz} - e^{-iz}\right],$$
$$\cos z = \tfrac{1}{2}\left[e^{iz} + e^{-iz}\right],$$
$$\sinh z = \tfrac{1}{2}\left[e^{z} - e^{-z}\right],$$
$$\cosh z = \tfrac{1}{2}\left[e^{z} + e^{-z}\right],$$

which are also analytic for all z.

7.3.2 SUMMARY

We now have four equivalent definitions of analyticity. A function

$$f(z) = u(x, y) + iv(x, y)$$

is analytic in a given region if any one of the following is true:

1. The derivative $f'(z)$ exists everywhere in the region.

2. The functions $u(x, y)$ and $v(x, y)$ have continuous derivatives and satisfy the Cauchy–Riemann relations.

3. The function $f(z)$ is continuous everywhere in the region and its integral around every closed contour in a simply connected part of the region is zero.

4. The function $f(z)$ can be represented by a power series expanded about any point in the region.

For example, consider the function

$$f(z) = z^2 + az + b. \tag{7.19}$$

From the definition in Eq. (7.5) we find the derivative to be

$$f'(z) = 2z + a,$$

which exists everywhere for $|z| < \infty$. Thus, $f(z)$ is analytic everywhere in the complex plane.

By writing $f(z)$ in the form of Eq. (7.4) the real and imaginary parts of $f(z)$ are seen to be, respectively,

$$u(x, y) = x^2 - y^2 + a_r x - a_i y + b_r \tag{7.20a}$$

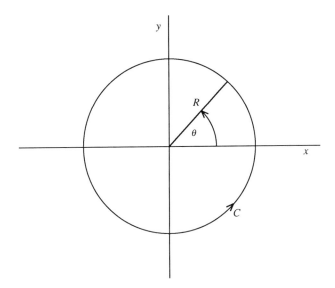

Figure 7.9

and
$$v(x, y) = 2xy + a_r y + a_i x + b_i. \tag{7.20b}$$

The subscripts r and i denote the real and imaginary parts of the complex numbers a and b. These functions have continuous derivatives and satisfy the Cauchy–Riemann equations.[5] Thus, $f(z)$ is analytic for all z.

For the integral
$$\oint_C f(z)\, dz = \oint_C (z^2 + az + b)\, dz,$$

we take the closed contour C to be a circle of radius R centered at the origin as shown in Figure 7.9. On C we have $z = Re^{i\theta}$. Then,
$$\oint_C f(z)\, dz = i \int_0^{2\pi} \left[R^3 e^{3i\theta} + aR^2 e^{2i\theta} + bRe^{i\theta} \right] d\theta,$$

which vanishes for all $R < \infty$. Since $f(z)$ is continuous everywhere, then it is analytic at every point in the complex plane.

Finally, we determine the higher derivatives of $f(z)$
$$f''(z) = 2,$$
$$f^{(n)}(z) = 0, \qquad n \geq 3$$

---

[5] See Exercise 7.9.

and construct the Taylor series of $f(z)$ about any point z_0 in the complex plane. That is,

$$f(z) = \sum_{n=0}^{\infty} \frac{[f^{(n)}(z)]_{z=z_0}}{n!} (z - z_0)^n$$

$$= (z_0^2 + az_0 + b) + (2z_0 + a)(z - z_0) + (z - z_0)^2,$$

which reduces to $f(z)$, as originally given in Eq. (7.19). Again the analyticity of $f(z)$ is established.

7.4   Laurent Expansion

If any function $f(z)$ has an isolated singularity at z_0, then it is analytic everywhere inside an annular ring bounded by two circles centered on z_0, as shown in Figure 7.10a. Let us make a cut line to connect the two circles and form a single closed contour

$$C = C_1 - C_2 + \text{cut};$$

see Figure 7.10b. Inside the ring we have from the Cauchy integral formula

$$f(z) = \frac{1}{2\pi i} \oint_C \frac{f(t)}{t - z} \, dt.$$

Because the contributions along the cut cancel, it follows that

$$f(z) = \frac{1}{2\pi i} \oint_{C_1} \frac{f(t)}{t - z} \, dt - \frac{1}{2\pi i} \oint_{C_2} \frac{f(t)}{t - z} \, dt. \tag{7.21}$$

We see from Figure 7.10b that on C_1 $|t - z_0| > |z - z_0|$. By following the procedure we used to obtain Eq. (7.16) from Eq. (7.15), we find that

$$\frac{1}{2\pi i} \oint_{C_1} \frac{f(t)}{t - z} \, dt = \sum_{k=0}^{\infty} \left(\frac{1}{2\pi i} \oint_{C_1} \frac{f(t)}{(t - z_0)^{k+1}} \, dt \right) (z - z_0)^k.$$

Also from Figure 7.10b it is clear that on C_2 $|t - z_0| < |z - z_0|$. In this case the above procedure leads to

$$\frac{1}{2\pi i} \oint_{C_2} \frac{f(t)}{t - z} \, dt = - \sum_{k=-1}^{-\infty} \left(\frac{1}{2\pi i} \oint_{C_2} \frac{f(t)}{(t - z_0)^{k+1}} \, dt \right) (z - z_0)^k.$$

On substituting these results into Eq. (7.21) we have

$$f(z) = \sum_{k=-\infty}^{\infty} a_k (z - z_0)^k \tag{7.22a}$$

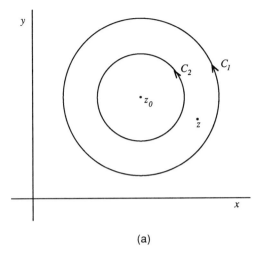

(a)

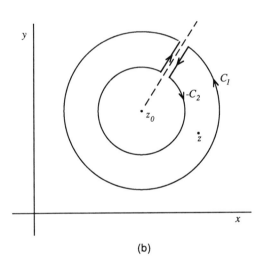

(b)

Figure 7.10

where

$$a_k = \frac{1}{2\pi i} \oint_C \frac{f(t)}{(t - z_0)^{k+1}} \, dt, \qquad k = 0, 1, 2, \ldots, \tag{7.22b}$$

and

$$a_k = \frac{1}{2\pi i} \oint_C \frac{f(t)}{(t - z_0)^{k+1}} \, dt, \qquad k = -1, -2, \ldots. \tag{7.22c}$$

This is the *Laurent series* for the function $f(z)$ in the annular region around z_0. We note that, because the integrand is analytic in the region between C_1 and C_2, we can use the same contour C for *both* integrals.

7.5   Essential Singularities

If $f(t)$ is analytic everywhere inside C (including z_0), then the integrand in Eq. (7.22c) is also analytic everywhere in this region. Hence, by the Cauchy integral theorem the coefficients a_k vanish for all $k < 0$ and the Laurent expansion (Eq. 7.22a), reduces to the Taylor expansion for $f(z)$. In this case the singularity is *removable*, because redefining $f(z)$ so that $f(z_0) = a_0$ does, in fact, remove the singularity.

For example, the function $f(z) = \sin z / z$ is not defined at $z = 0$. However, the Taylor series of $\sin z/z$ is

$$\frac{\sin z}{z} = 1 - \frac{z^2}{3!} + \frac{z^4}{5!} - \frac{z^6}{7!} + \cdots ,$$

which is clearly equal to one if $z = 0$. Thus, we define

$$f(0) = \lim_{z \to 0} \frac{\sin z}{z} = 1,$$

and the singularity of $f(z)$ at $z = 0$ is removed.

Suppose a function $f(z)$ contains only a finite number of negative powers of $z - z_0$,

$$f(z) = \frac{a_{-m}}{(z - z_0)^m} + \frac{a_{-m+1}}{(z - z_0)^{m-1}} + \cdots + \frac{a_{-1}}{z - z_0} + \sum_{k=0}^{\infty} a_k (z - z_0)^k .$$

If $a_{-m} \neq 0$, then $f(z)$ is said to have a *pole of order m* at z_0 (see Section 2.3.2).

An isolated singularity that is neither removable nor a pole is called an *essential singularity*.[6]

From Eq. (7.22c) we see that

$$a_{-1} = \frac{1}{2\pi i} \oint_C f(t)\, dt. \tag{7.23}$$

If $f(t)$ is analytic everywhere inside C (including z_0), then $a_{-1} = 0$ by the Cauchy integral theorem. However, if $f(t)$ has an isolated singularity inside C at z_0, then, in general, $a_{-1} \neq 0$. The coefficient a_{-1} is called the *residue* of $f(z)$ at z_0. This turns out to be an extremely fruitful notion as we shall see in the next chapter.

---

[6]See p. 25 for an example.

7.6   Branch Points

The function $z^{1/2}$ is *not* analytic at $z = 0$, since its derivative does not exist there. This is an example of a special kind of singularity called a *branch point*. Its peculiarity arises from the multivaluedness of the function. In polar notation $z = re^{i\theta}$ and

$$z^{1/2} = r^{1/2}e^{i\theta/2}.$$

Now consider the variation of $z^{1/2}$ as we proceed around a circle of radius r about the origin. The path is depicted in Figure 7.11a. If we start on the x axis with $\theta = 0$, we have

$$z^{1/2}\big|_{\theta=0} = r^{1/2}.$$

As θ approaches 2π, we approach this same point on the x axis, but the function now has the value

$$z^{1/2}\big|_{\theta=2\pi} = -r^{1/2}.$$

It has two *different values* at the *same point* in the complex plane. The function is discontinuous across any cut extending out from $z = 0$ (in this case the positive real axis).

Clearly, the trouble with the singularity at $z = 0$ is not at the point itself, but in the *neighborhood* of the singularity. The singularity is *not isolated*. We can eliminate this problem and make $z^{1/2}$ single valued by (for example) restricting θ to values less than 2π. That is,

$$z^{1/2} = r^{1/2}e^{i\theta/2}, \qquad 0 \le \theta < 2\pi.$$

This defines a particular *branch* of $z^{1/2}$. Another branch is defined over the interval $2\pi \le \theta < 4\pi$. We make a *branch cut* along the positive real axis that excludes the discontinuity on this axis as seen in Figure 7.11b. The cut may be along *any* line which extends in to the branch point. In the example here we could have chosen the cut along the *negative* real axis in which case $z^{1/2}$ would be single valued for $-\pi < \theta \le \pi$.

Notice that a branch cut differs from the cut line introduced to transform a multiply connected region into a simply connected one. In this latter case the function is *continuous* across the cut line. We shall have to handle branch points with particular care.

7.7   Analytic Continuation

We may have an analytic function that is defined in a given region by a series, an integral, or some other mathematical expression. Outside the

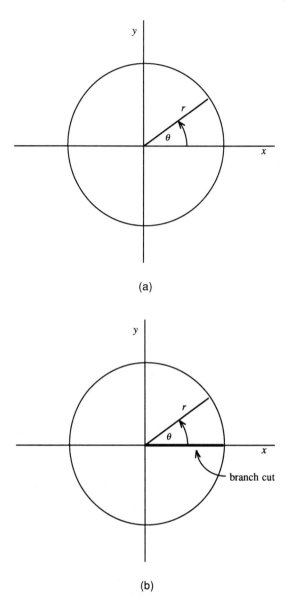

(a)

(b)

Figure 7.11

region the expression may be meaningless, but it may still be possible for the function itself to have a meaning there.

For example, the geometric series

$$u(z) = \sum_{k=0}^{\infty} z^k \tag{7.24}$$

represents an analytic function $u(z)$ everywhere inside a circle of unit radius centered at the origin of the complex plane. Outside this circle the series diverges and cannot represent an analytic function.

From Eq. (2.6) we see that the Taylor expansion of the function

$$v(z) = \frac{1}{1-z} \tag{7.25}$$

is just the geometric series in z. Now, $v(z)$ is analytic *everywhere* in the complex plane except on the real axis at $z = 1$. Clearly, its domain of validity extends beyond that for which $u(z)$ is defined.

We regard $u(z)$ and $v(z)$ as being the same function, but the representation in Eq. (7.24) has a much smaller domain of validity than the representation in Eq. (7.25). Where a function $v(z)$ coincides with an analytic function $u(z)$ over some region for which $u(z)$ is defined and is analytic beyond that region, that function is said to be an *analytic continuation* of $u(z)$.

Let us consider an analytic function $u(z)$ whose exact form is not known to us, but whose Taylor series is

$$u(z) = \sum_{k=0}^{\infty} a_k^{(1)}(z - z_1)^k$$

for points inside a circle C_1 centered at z_1 as shown in Figure 7.12a. Now choose a point z_2 inside C_1 and evaluate $u(z)$ and its derivatives at z_2. From these, we can in principle construct a series to represent $u(z)$ inside a circle C_2 centered at z_2 (as in Figure 7.12b) provided there are no singularities inside C_2, particularly at points on C_1 that lie inside C_2.

By the process of analytic continuation we have extended the domain of validity of $u(z)$ beyond the original region inside C. The procedure can be continued indefinitely by choosing inside each succeeding circle C_{n-1} a new point z_n about which to construct a new Taylor series for $u(z)$ inside a new circle C_n as illustrated in Figure 7.12c. Some examples follow.

The function

$$f(z) = \frac{1}{n!} \int_0^{\infty} t^n e^{-zt} dt \quad \text{with} \quad n = 0, 1, 2, \ldots$$

is analytic for $\text{Re}(z) > 0$. From Eq. (1.16) we see that

$$f(z) = \frac{1}{z^{n+1}} \quad \text{for} \quad \text{Re}(z) > 0.$$

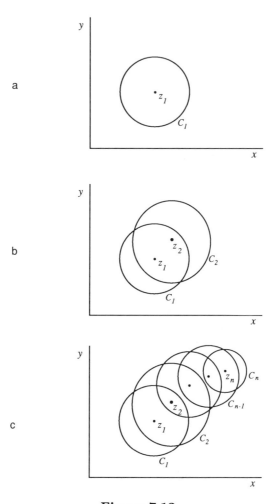

Figure 7.12

The function z^{-n-1} is analytic everywhere except at $z = 0$. Thus, z^{-n-1} is the analytic continuation of $f(z)$ into the half-plane $\text{Re}(z) < 0$.

As a second example we show that the two series

$$u(z) = 2\sum_{k=0}^{\infty} \frac{(4z - 3i)^k}{(4 - 3i)^{k+1}} \quad \text{and} \quad v(z) = \sum_{k=0}^{\infty} \frac{(2z + 1)^k}{3^{k+1}}$$

are analytic continuations of each other. Both of these can be written as

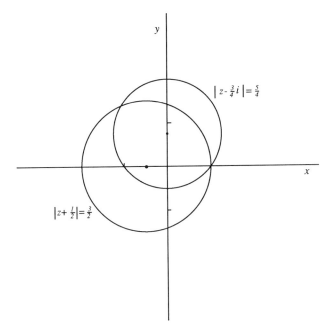

Figure 7.13

geometric series. We have

$$u(z) = \frac{2}{4-3i} \sum_{k=0}^{\infty} \left(\frac{4z-3i}{4-3i} \right)^k = \frac{2/(4-3i)}{1-(4z-3i)/(4-3i)},$$

which converges for $\left| \frac{4z-3i}{4-3i} \right| < 1$ or $\left| z - \frac{3i}{4} \right| < \frac{5}{4}$.
    Similarly,

$$v(z) = \frac{1}{3} \sum_{k=0}^{\infty} \left(\frac{2z+1}{3} \right)^k = \frac{1/3}{1-(2z+1)/3}$$

converges for $\left| \frac{2z+1}{3} \right| < 1$ or $\left| z + \frac{1}{2} \right| < \frac{3}{2}$.
    The circles of convergence for these series are shown in Figure 7.13. The two series represent the same function $(2-2z)^{-1}$ in the region common to both circles. Therefore, they are analytic continuations of each other.

EXERCISES

1. Carry out the indicated operations and write out explicitly the real and imaginary parts of each of the results:

   a. $(3 - 2i) + (5i - 2)$.

   b. i^9.

   c. $i^7 + i^4$.

   d. $\left(\frac{2+i}{1-i}\right)^2 - 2\left(\frac{1-i}{1+i}\right)^3$.

   e. $(2 - i)(3 + 4i)(1 - 5i)$.

   f. $3(2i - 1) - 5(4 - 3i)$.

2. Write each of the following complex numbers explicitly in the form $a + ib$ where a and b are real numbers:

   a. $\frac{1+3i}{1-i}$.

   b. $\sqrt{1 + \sqrt{3}i}$.

   c. $\sqrt{i}$.

   d. $\frac{7-i}{3-4i}$.

   e. $\cos\left(\frac{\pi}{4} + ib\right)$.

   f. $\sqrt{8 + 3i\sqrt{2i}}$.

3. Let r and y be real. Show that r^{iy} is complex with unit amplitude. Write the real and imaginary parts of r^{iy} explicitly in terms of r and y.

4. Show that for real x

$$\cos ix = \cosh x \quad \text{and} \quad \sin ix = i \sinh x.$$

5. Show that

$$_1F_1(1; 2; 2ix) = e^{ix} \frac{\sin x}{x}.$$

6. Write each of the following functions explicitly in the form of Eq. (7.4),

$$f(z) = u(x, y) + iv(x, y)$$

and show that the Cauchy–Riemann equations hold for each function:

   a. $f(z) = ze^{-z}$.

   b. $f(z) = (1 + z)^{-1}$ for $z \neq -1$.

   c. $f(z) = \sin z$.

7. In the integral

$$\oint_C \frac{e^z}{z-a}\, dz$$

a is a real, positive constant. Use the Cauchy integral formula to evaluate this integral for the two cases:

   a. C is the circle $|z| = 2a$.

   b. C is the circle $|z| = \frac{1}{2}a$.

8. Evaluate the integral

$$\oint \frac{dz}{z}$$

along each of the two paths shown in Figure 7.14. For $C = C_1$ let $z = x + iy$ and show that the integral reduces to

$$\oint_{C_1} \frac{dz}{z} = 4i \int_{-1}^{1} \frac{dt}{t^2 + 1}.$$

Make a variable change $t = \tan\phi$ to evaluate this latter integral. With $C = C_2$ take $z = e^{i\theta}$. Since the value of the integral is independent of the path, the results for both paths should be the same.

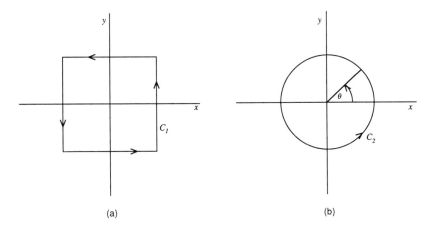

(a)   (b)

Figure 7.14

9. Complete the details of the calculations in the four different ways of demonstrating the analyticity of the function

$$f(z) = z^2 + az + b$$

given in Eq. (7.19).

10. Show that the function e^z defined by

$$e^z = e^x(\cos y + i \sin y)$$

is analytic for all $z = x + iy$.

11. For any function $F(x + iy)$ where x and y are real, show that

$$F(x - iy) = e^{-2i\eta} F(x + iy)$$

with $\eta = \arg F(x + iy)$.

12. The function $f(z)$ defined by the series

$$f(z) = \sum_{k=1}^{\infty} \frac{(-1)^{k+1} z^k}{k}$$

is analytic for $|z| < 1$. Find the analytic function that is the continuation of $f(z)$ into the entire complex plane except for the point $z = -1$. **Hint**: Differentiate term by term and identify the resulting series.

13. For the two geometric series

$$u(z) = 2 \sum_{k=0}^{\infty} \frac{(2z + i)^k}{(4 + i)^{k+1}} \quad \text{and} \quad v(z) = \sum_{k=0}^{\infty} \frac{(z - i)^k}{(2 - i)^{k+1}},$$

plot the circle of convergence for each on the same graph. Show that they represent the same function in their respective regions of convergence and are analytic continuations of each other.

14. Prove that the two series

$$f_1(z) = \sum_{k=0}^{\infty} \frac{(2z)^k}{9^{k+1}} \quad \text{and} \quad f_2(z) = \sum_{k=0}^{\infty} \frac{(2z - 1 - 6i)^k}{(8 - 6i)^{k+1}}$$

are representations of the same function $f(z)$, which has the form

$$f(z) = \frac{1}{a - bz}.$$

Show on the same graph the regions of convergence of the series and show that they are analytic continuations of each other.

8

Applications of Contour Integrals

The notion of a residue, introduced in Chapter 7, provides a useful tool to evaluate definite integrals, as shown below. First, let us extend the idea to a multiply connected region in the complex plane, that is, one with several holes or places where the function $f(z)$ is not analytic.

8.1 The Cauchy Residue Theorem

Consider a region with two holes, as in Figure 8.1a. The region inside C and outside C_1 and C_2 contains no holes and is therefore simply connected. Now let us make cut lines as in Figure 8.1b so that the single continuous contour

$$C' = C + (-C_1) + (-C_2) + \text{cut lines}$$

encloses a simply connected region. We have

$$\oint_{C'} f(z)\, dz = \oint_C f(z)\, dz + \oint_{-C_1} f(z)\, dz + \oint_{-C_2} f(z)\, dz.$$

The left-hand side vanishes by Cauchy's theorem. This gives

$$\oint_C f(z)\, dz = \oint_{C_1} f(z)\, dz + \oint_{C_2} f(z)\, dz.$$

This result is easily extended to any number of holes, say N holes. Then,

$$\oint_C f(z)\, dz = \sum_{k=1}^{N} \oint_{C_k} f(z)\, dz,$$

where C surrounds all holes while C_k surrounds only the kth hole. Now let C enclose N isolated singularities of $f(z)$, the kth one being at z_k. Then from Eq. (7.23) we have

$$\oint_C f(z)\, dz = \sum_{k=1}^{N} \oint_{C_k} f(z)\, dz = 2\pi i \sum_{k=1}^{N} a_{-1}(z_k)$$

$$= 2\pi i [\text{sum of residues}]. \tag{8.1}$$

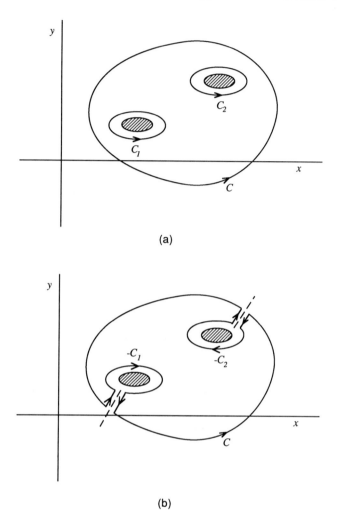

(a)

(b)

Figure 8.1

This is the *Cauchy residue theorem*.

If we have a means of determining the coefficient a_{-1} in the Laurent expansion of $f(z)$ about an isolated singularity at z_0, then this theorem can be used to evaluate integrals around closed contours. Suppose $f(z)$ has a pole of order n at $z = z_0$. Then,

$$f(z) = \frac{a_{-n}}{(z - z_0)^n} + \frac{a_{-n+1}}{(z - z_0)^{n-1}} + \cdots + \frac{a_{-1}}{(z - z_0)} + \sum_{k=0}^{\infty} a_k(z - z_0)^k \quad (8.2)$$

where $a_{-n} \neq 0$.

For $n = 1$ multiply this equation by $z - z_0$ and take the limit $z \to z_0$ to get

$$\lim_{z \to z_0} (z - z_0)f(z) = \lim_{z \to z_0} \left[a_{-1} + \sum_{k=0}^{\infty} a_k(z - z_0)^{k+1} \right] = a_{-1}.$$

For $n = 2$ multiply Eq. (8.2) by $(z - z_0)^2$ to obtain

$$(z - z_0)^2 f(z) = a_{-2} + (z - z_0)a_{-1} + \sum_{k=0}^{\infty} a_k(z - z_0)^{k+2}.$$

Now differentiate with respect to z and take the limit $z \to z_0$. This gives

$$\lim_{z \to z_0} \frac{d}{dz} [(z - z_0)^2 f(z)] = a_{-1}.$$

Similarly, for arbitrary n we multiply Eq. (8.2) by $(z - z_0)^n$ and differentiate $n - 1$ times to get

$$\frac{d^{n-1}}{dz^{n-1}} [(z - z_0)^n f(z)] = 1 \cdot 2 \cdot 3 \cdots (n - 1)a_{-1}$$

$$+ \sum_{k=0}^{\infty} (k + n)(k + n - 1) \cdots (k + 2)a_k(z - z_0)^{k+1}.$$

Taking the limit $z \to z_0$ again, we find that

$$a_{-1} = \frac{1}{(n - 1)!} \lim_{z \to z_0} \left[\frac{d^{n-1}}{dz^{n-1}} ((z - z_0)^n f(z)) \right], \qquad (8.3)$$

which is the formula for the residue of $f(z)$ for the nth order pole at $z = z_0$.
    Consider the integral

$$\oint_C \frac{dz}{z - 1}, \qquad (8.4)$$

where C is any contour that encloses the point on the real axis at $z = 1$ as shown in Figure 8.2a. We can evaluate this integral directly by choosing C to be a circle of radius r centered at $z = 1$, as shown in Figure 8.2b. Write z in polar representation as

$$z = 1 + re^{i\theta}.$$

By direct substitution into Eq. (8.4) this gives

$$\oint_C \frac{dz}{z - 1} = \oint_0^{2\pi} i d\theta = 2\pi i. \qquad (8.5)$$

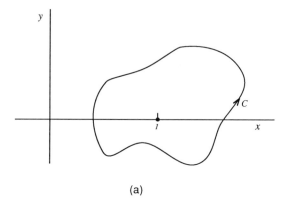

(a)

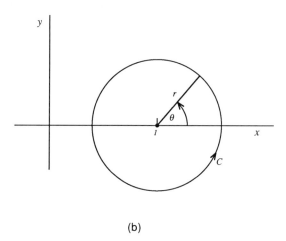

(b)

Figure 8.2

Now apply the residue theorem to the evaluation of this integral. Because it has a first-order pole at $z = 1$, we have

$$\oint_C \frac{dz}{z-1} = 2\pi i a_{-1}$$

$$= 2\pi i \lim_{z \to 1} \left[(z-1) \frac{1}{z-1} \right]$$

$$= 2\pi i,$$

which agrees with Eq. (8.5). Note that it was not necessary here to specify the exact shape of the contour C.

8.2   Evaluation of Definite Integrals by Contour Integration

The residue theorem is particularly useful for evaluating certain kinds of definite integrals. We demonstrate the technique by several examples.

In evaluating integrals of the type

$$\int_{-\infty}^{\infty} f(x)\,dx,$$

we choose a contour C, as in Figure 8.3, such that part of it C_1 is along the real axis and such that the integral along the remaining part C_2 is either zero or simple to evaluate. Then,

$$\int_{-\infty}^{\infty} f(x)\,dx = \lim_{R\to\infty} \int_{-R}^{R} f(z)\,dz$$

$$= \lim_{R\to\infty} \left[\oint_C f(z)\,dz - \int_{C_2} f(z)\,dz \right]$$

$$= 2\pi i[\text{sum of residues in } C] - \lim_{R\to\infty} \int_{C_2} f(z)\,dz.$$

Example 1. The function

$$f(z) = \frac{1}{z^2 + 1}$$

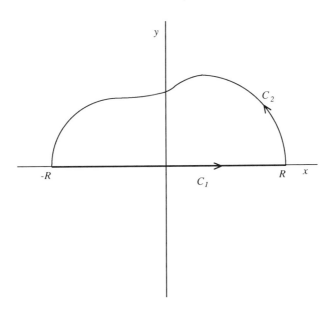

Figure 8.3

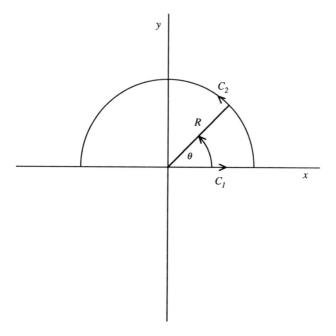

Figure 8.4

clearly has first-order poles at $z = \pm i$. Choose for C_2 the semicircle of radius R centered at the origin and lying in the upper half-plane as in Figure 8.4. The complete contour $C = C_1 + C_2$ encloses the pole at $z = i$. Thus,

$$\int_{-\infty}^{\infty} \frac{dx}{x^2 + 1} = \lim_{R \to \infty} \left[\oint_C \frac{dz}{z^2 + 1} - \int_{C_2} \frac{dz}{z^2 + 1} \right]$$

$$= \lim_{R \to \infty} \left[2\pi i a_{-1} - \int_0^{\pi} \frac{iRe^{i\theta}\, d\theta}{R^2 e^{2i\theta} + 1} \right].$$

The second term in the square brackets vanishes in the limit $R \to \infty$. From Eq. (8.3) with $n = 1$ we find

$$a_{-1} = \lim_{z \to i} \left[(z - i) \frac{1}{(z - i)(z + i)} \right] = \frac{1}{2i}.$$

For the value of the integral we have

$$\int_{-\infty}^{\infty} \frac{dx}{x^2 + 1} = \pi.$$

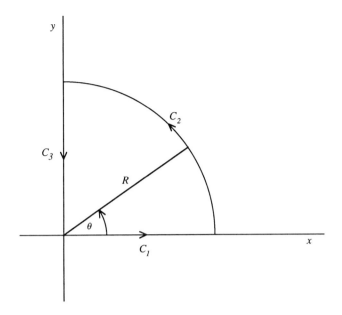

Figure 8.5

Note that the integrand is an even function of x so that

$$\int_0^\infty \frac{dx}{x^2 + 1} = \frac{\pi}{2}.$$

Example 2. To evaluate the integral

$$\int_0^\infty \frac{dx}{x^4 + 1}$$

choose a contour consisting of the three pieces shown in Figure 8.5. Then

$$\int_0^\infty \frac{dx}{x^4 + 1} = \lim_{R\to\infty} \int_0^R \frac{dx}{x^4 + 1}$$

$$= \lim_{R\to\infty} \left[\oint_C \frac{dz}{z^4 + 1} - \int_{C_2} \frac{dz}{z^4 + 1} - \int_{C_3} \frac{dz}{z^4 + 1} \right]$$

$$= \lim_{R\to\infty} \left[\oint_C \frac{dz}{z^4 + 1} - \int_0^{\pi/2} \frac{iRe^{i\theta}\, d\theta}{R^4 e^{4i\theta} + 1} - \int_R^0 \frac{i\,dy}{(iy)^4 + 1} \right].$$

The second term in square brackets vanishes in the limit $R \to \infty$. So we have

$$\int_0^\infty \frac{dx}{x^4 + 1} = \lim_{R\to\infty} \oint_C \frac{dz}{z^4 + 1} - i \int_\infty^0 \frac{dy}{y^4 + 1}.$$

Because y is a dummy variable of integration, we can set $y = x$ and interchange the limits of integration to get

$$(1 - i) \int_0^\infty \frac{dx}{x^4 + 1} = \lim_{R \to \infty} \oint_C \frac{dz}{z^4 + 1}$$

$$= \lim_{R \to \infty} [2\pi i(\text{sum of residues})].$$

We see that

$$z^4 + 1 = \left(z - e^{i\pi/4}\right)\left(z - e^{i3\pi/4}\right)\left(z - e^{i5\pi/4}\right)\left(z - e^{i7\pi/4}\right).$$

The contour C encloses the simple pole at $z = e^{i\pi/4}$. The residue is

$$a_{-1} = \frac{-i(1 - i)}{4\sqrt{2}}.$$

Thus,

$$\int_0^\infty \frac{dx}{x^4 + 1} = \frac{\pi}{2\sqrt{2}}.$$

Example 3. This example requires the calculation of residues of higher order poles. We use the residue theorem to evaluate the integral

$$I = \int_0^\infty \frac{dx}{(x^2 + 1)^2}.$$

Since the integrand is an even function of x,

$$I = \frac{1}{2} \int_{-\infty}^\infty \frac{dx}{(x^2 + 1)^2}$$

$$= \frac{1}{2} \lim_{R \to \infty} \left[\oint_C \frac{dz}{(z^2 + 1)^2} - i \int_0^\pi \frac{Re^{i\theta} d\theta}{(R^2 e^{2i\theta} + 1)^2} \right].$$

The closed curve C consists of the portion of the real axis from $-R$ to R and a semicircle of radius R in the upper half plane as shown in Figure 8.4. The second integral in the square brackets vanishes in the limit. The contour C encloses a second-order pole on the imaginary axis at $z = i$. Evaluating the residue according to Eq. (8.3) yields

$$I = \pi i \lim_{z \to i} \left[\frac{-2}{(z + i)^3} \right]$$

or

$$\int_0^\infty \frac{dx}{(x^2 + 1)^2} = \frac{\pi}{4}.$$

Example 4. Another class of definite integrals may be evaluated by means of contour integration and involves trigonometric functions. For example, if in the integral

$$\int_0^{2\pi} \frac{d\theta}{a + b\cos\theta} \quad \text{with} \quad a > b > 0$$

the variable of integration is changed to $z = e^{i\theta}$, z traces out a closed path on a unit circle in the complex plane as θ goes from 0 to 2π. Thus,

$$\int_0^{2\pi} \frac{d\theta}{a + b\cos\theta} = -2i \oint_C \frac{dz}{bz^2 + 2az + b},$$

where C is a circle of radius 1 centered at the origin. The integrand on the right-hand side has simple poles at

$$z = -\frac{a}{b} \pm \sqrt{\frac{a^2}{b^2} - 1}.$$

With $a > b > 0$, only the pole corresponding to the upper sign lies inside the unit circle. Applying the residue theorem we obtain

$$\int_0^{2\pi} \frac{d\theta}{a + b\cos\theta} = \frac{2\pi}{\sqrt{a^2 - b^2}}.$$

In general, we may use this technique to evaluate integrals of the form

$$I = \int_0^{2\pi} F(\sin\theta, \cos\theta)\, d\theta$$

where $F(\sin\theta, \cos\theta)$ is a rational function of $\sin\theta$ and $\cos\theta$. Set $z = e^{i\theta}$ and make the substitutions,

$$\cos\theta = \tfrac{1}{2}(z + z^{-1}),$$
$$\sin\theta = -\tfrac{1}{2}i(z - z^{-1}),$$
$$d\theta = -iz^{-1}dz,$$

to obtain

$$I = -\oint_C F\left(\frac{z^2 - 1}{2iz}, \frac{z^2 + 1}{2z}\right) \frac{dz}{z}$$
$$= 2\pi[\text{sum of residues of } F/z \text{ inside } C],$$

where C is the unit circle centered at the origin.

8.2.1 JORDAN'S LEMMA

Example 5. Next, consider the integral

$$\oint_C \frac{e^{iaz}}{1 + z^2}\, dz = \int_{C_1} \frac{e^{iaz}}{1 + z^2}\, dz + \int_{C_2} \frac{e^{iaz}}{1 + z^2}\, dz \quad \text{for} \quad \text{Re}(a) \geq 0,$$

where the contour $C = C_1 + C_2$ is depicted in Figure 8.4. If we can show that the integral over C_2 makes no contribution in the limit $R \to \infty$, then

$$\int_{-\infty}^{\infty} \frac{e^{iax}}{1 + x^2}\, dx = \lim_{R \to \infty} \oint_C \frac{e^{iaz}}{1 + z^2}\, dz. \tag{8.6}$$

To establish this result we take $z = Re^{i\theta} = R(\cos\theta + i\sin\theta)$ in the integral over C_2,

$$\int_{C_2} \frac{e^{iaz}}{1+z^2}\,dz = i\int_0^\pi \frac{Re^{-aR\sin\theta}}{1+R^2e^{2i\theta}}\,e^{i(\theta+aR\cos\theta)}\,d\theta \tag{8.7}$$

which clearly vanishes in the limit $R \to \infty$. The contour in Eq. (8.6) encloses only the singularity at $z = +i$ for which the residue is $-\frac{1}{2}ie^{-a}$. Thus,

$$\int_{-\infty}^\infty \frac{e^{iax}}{1+x^2}\,dx = \pi e^{-a}. \tag{8.8}$$

To take a slightly different approach we first note from Eq. (8.7) that

$$\left|\int_{C_2} \frac{e^{iaz}}{1+z^2}\,dz\right| \le \frac{R}{|1-R^2|}\int_0^\pi e^{-aR\sin\theta}\,d\theta. \tag{8.9}$$

Now write the integral on the right-hand side as the sum of two integrals; one for $0 \le \theta \le \pi/2$ and the other for $\pi/2 \le \theta \le \pi$. In the latter one, replace θ by $\pi - \theta$ as the variable of integration. This gives

$$\int_0^\pi e^{-aR\sin\theta}\,d\theta = 2\int_0^{\pi/2} e^{-aR\sin\theta}\,d\theta.$$

Note that

$$\frac{d}{d\theta}\left(\frac{\sin\theta}{\theta}\right) = \frac{\cos\theta}{\theta^2}(\theta - \tan\theta). \tag{8.10}$$

In constructing the Taylor series of $\tan\theta$ we find that the kth derivative of $\tan\theta$ is a sum of even or odd powers of $\tan\theta$ if k is odd or even, respectively. For $\theta = 0$, only the zeroth-power term is nonvanishing. Thus, the Taylor expansion of $\tan\theta$ about $\theta = 0$ is an infinite sum of odd powers of θ with *positive* coefficients. The first few of the terms are given in Eq. (2.23). Clearly,

$$\theta < \tan\theta \quad \text{for} \quad 0 < \theta < \frac{\pi}{2}.$$

Therefore, Eq. (8.10) shows that in the interval $0 < \theta < \pi/2$, $\frac{\sin\theta}{\theta}$ is a *decreasing* function of θ. Because this ratio has its maximum value at $\theta = 0$ where it is equal to one,

$$\frac{\sin\theta}{\theta} \ge \frac{\sin(\pi/2)}{\pi/2} \quad \text{for} \quad 0 \le \theta \le \pi/2,$$

or

$$\sin\theta \ge \frac{2\theta}{\pi} \quad \text{for} \quad 0 \le \theta \le \pi/2. \tag{8.11}$$

This relation can also be seen by comparing the graphs of $\sin\theta$ and $2\theta/\pi$ as functions of θ. Thus,

$$\int_0^\pi e^{-aR\sin\theta}\,d\theta \le 2\int_0^{\pi/2} e^{-2aR\theta/\pi}\,d\theta = \frac{\pi}{aR}(1 - e^{-aR}). \tag{8.12}$$

We use this result in Eq. (8.9) to show that

$$\left| \int_{C_2} \frac{e^{iaz}}{1+z^2} \, dz \right| \le \frac{\pi(1 - e^{-aR})}{a|R^2 - 1|} \xrightarrow[R \to \infty]{} 0.$$

Again, Eq. (8.6) is established.

This procedure is seen to apply to *any* integral of the form

$$\int_{-\infty}^{\infty} e^{iax} f(x) \, dx$$

provided $f(z)$ satisfies certain conditions which are not very stringent. This integral can be written as

$$\int_{-\infty}^{\infty} e^{iax} f(x) \, dx = \lim_{R \to \infty} \left[\oint_C e^{iaz} f(z) \, dz - \int_{C_2} e^{iaz} f(z) \, dz \right], \qquad (8.13)$$

with the contour C defined in Figure 8.4. On the semicircle C_2 we have $z = Re^{i\theta}$. Thus, for the integral over C_2,

$$\left| \int_{C_2} e^{iaz} f(z) \, dz \right| \le \left| R \int_0^\pi f(Re^{i\theta}) e^{-aR \sin \theta} \, d\theta \right|.$$

If for all θ in the interval $0 \le \theta \le \pi$

$$|f(Re^{i\theta})| \le g(R)$$

where $g(R)$ is some real function of R, then

$$\left| \int_{C_2} e^{iaz} f(z) \, dz \right| \le \frac{\pi}{a} g(R)(1 - e^{-aR}).$$

Again, we have used Eq. (8.12).

Therefore, if we invoke the condition

$$g(R) \xrightarrow[R \to \infty]{} 0,$$

then the contribution along C_2 in Eq. (8.13) vanishes and

$$\int_{-\infty}^{\infty} e^{iax} f(x) \, dx = 2\pi i \left[\begin{array}{l} \text{sum of residues of } e^{iaz} f(z) \\ \text{in the upper half plane} \end{array} \right]. \qquad (8.14a)$$

This is *Jordan's lemma*.

A similar result may be obtained for

$$\int_{-\infty}^{\infty} e^{-iaz} f(x) \, dx$$

by completing the contour in the lower half plane. If $|f(Re^{-i\theta})| \leq g(R)$, where $g(R) \xrightarrow[R\to\infty]{} 0$ for all θ in the interval $\pi \leq \theta \leq 2\pi$, then

$$\int_{-\infty}^{\infty} e^{-iax} f(x)\, dx = -2\pi i \begin{bmatrix} \text{sum of residues of } e^{-iaz} f(z) \\ \text{in the lower half plane} \end{bmatrix}. \qquad (8.14b)$$

Example 6. Evaluate the integral

$$\int_0^\infty \frac{x \sin x}{a^2 + x^2}\, dx.$$

The integrand is an even function of x; therefore, we can write

$$\int_0^\infty \frac{x \sin x}{a^2 + x^2}\, dx = \frac{1}{4i} \left[\int_{-\infty}^{\infty} \frac{x e^{ix}}{a^2 + x^2}\, dx - \int_{-\infty}^{\infty} \frac{x e^{-ix}}{a^2 + x^2}\, dx \right]. \qquad (8.15)$$

The integrand of the first integral on the right-hand side is of the form $f(z)e^{iz}$, where

$$|f(z)| = |f(Re^{i\theta})| = \left| \frac{Re^{i\theta}}{4i(a^2 + R^2 e^{2i\theta})} \right| \leq \frac{R}{4|R^2 - a^2|}. \qquad (8.16)$$

With $g(R) = R/(4(R^2 - a^2))$ it is clear that the condition $g(R) \xrightarrow[R\to\infty]{} 0$ for Jordan's lemma is satisfied by the first integral on the right-hand side of Eq. (8.15). By completing the contour in the upper half plane we enclose the simple pole at $z = ia$ with residue $a_{-1} = \frac{1}{2}e^{-a}$. Hence,

$$\int_{-\infty}^{\infty} \frac{x e^{ix}}{a^2 + x^2}\, dx = i\pi e^{-a}. \qquad (8.17)$$

A result similar to Eq. (8.17) also is obtained for the second integral in Eq. (8.15). Completing the contour in the *lower* half plane yields

$$\int_{-\infty}^{\infty} \frac{x e^{-ix}}{a^2 + x^2}\, dx = -i\pi e^{-a}.$$

Thus,

$$\int_0^\infty \frac{x \sin x}{a^2 + x^2}\, dx = \frac{\pi}{2} e^{-a}.$$

Example 7. Here, we use contour integration to evaluate the integrals[1]

$$\int_0^\infty \cos x^\nu dx \quad \text{and} \quad \int_0^\infty \sin x^\nu dx,$$

---

[1]With $\nu = 2$ these integrals are related to the Fresnel integrals, which occur in the analysis of Fresnel diffraction in optics. See also Exercise 10.6.

where ν is real and $\nu > 1$. The values of both integrals can be determined simultaneously from the integral

$$I_\nu = \oint_C e^{it^\nu} dt.$$

The contour C consists of the three pieces shown in Figure 8.6. Since the integrand is analytic everywhere in the complex plane, I_ν vanishes. Carrying out the integration explicitly over the three pieces yields

$$I_\nu = \int_0^R e^{ix^\nu} dx + iR \int_0^{\pi/2\nu} e^{iR^\nu e^{i\nu\theta}} e^{i\theta} d\theta + e^{i\pi/2\nu} \int_R^0 e^{ir^\nu e^{i\pi/2}} dr = 0. \tag{8.18}$$

For the second integral

$$\left| iR \int_0^{\pi/2\nu} e^{i(\theta + R^\nu \cos\nu\theta)} e^{-R^\nu \sin\nu\theta} d\theta \right| \le R \int_0^{\pi/2\nu} e^{-R^\nu \sin\nu\theta} d\theta$$

$$= \frac{R}{\nu} \int_0^{\pi/2} e^{-R^\nu \sin\phi} d\phi$$

$$\le \frac{R}{\nu} \int_0^{\pi/2} e^{-2R^\nu \phi/\pi} d\phi$$

$$= \frac{\pi}{2\nu R^{\nu-1}} \left(1 - e^{-R^\nu}\right) \underset{R\to\infty}{\longrightarrow} 0.$$

Thus, if we take the limit $R \to \infty$ in Eq. (8.18), we find that

$$\int_0^\infty e^{ix^\nu} dx = e^{i\pi/2\nu} \int_0^\infty e^{-r^\nu} dr. \tag{8.19}$$

If the variable of integration is changed to $u = r^\nu$, the integral on the right-hand side becomes

$$\int_0^\infty e^{-r^\nu} dr = \frac{1}{\nu} \Gamma\left(\frac{1}{\nu}\right).$$

With this result Eq. (8.19) may be rewritten

$$\int_0^\infty \left(\cos x^\nu + i \sin x^\nu\right) dx = \left(\cos\frac{\pi}{2\nu} + i \sin\frac{\pi}{2\nu}\right) \frac{1}{\nu} \Gamma\left(\frac{1}{\nu}\right).$$

Finally, equate the real and imaginary parts separately on the two sides of this equation to get

$$\int_0^\infty \cos x^\nu dx = \frac{1}{\nu} \Gamma\left(\frac{1}{\nu}\right) \cos\frac{\pi}{2\nu}, \tag{8.20}$$

and

$$\int_0^\infty \sin x^\nu dx = \frac{1}{\nu} \Gamma\left(\frac{1}{\nu}\right) \sin\frac{\pi}{2\nu}. \tag{8.21}$$

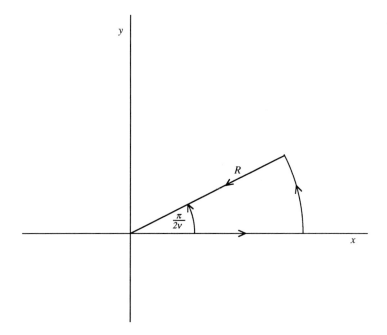

Figure 8.6

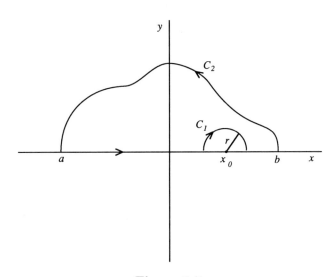

Figure 8.7

8.2.2 CAUCHY PRINCIPAL VALUE

Let $f(z)$ be a function with a simple pole on the real axis at $z = x_0$. If x_0 is in the interval $a < x_0 < b$, the integral

$$\int_a^b f(x)\,dx$$

does not strictly exist. The *Cauchy principal value* of this integral is defined as the limit

$$P \int_a^b f(x)\,dx = \lim_{r \to 0} \left[\int_a^{x_0-r} f(x)\,dx + \int_{x_0+r}^b f(x)\,dx \right]. \tag{8.22}$$

We evaluate this integral by completing the contour in the upper half-plane as shown in Figure 8.7. The small semicircle C_1 excludes the singularity at x_0. The integral over the closed contour is in four pieces,

$$\oint_C f(z)\,dz = \int_a^{x_0-r} f(x)\,dx + \int_{C_1} f(z)\,dz$$

$$+ \int_{x_0+r}^b f(x)\,dx + \int_{C_2} f(z)\,dz. \tag{8.23}$$

In the limit $r \to 0$ the first and third terms on the right hand side yield the Cauchy principal value. Expand $f(z)$ in a small neighborhood around the point x_0 where $f(z)$ has a simple pole, that is,

$$f(z) = \frac{a_{-1}}{z - x_0} + \sum_{k=0}^{\infty} a_k (z - x_0)^k.$$

Because $z = x_0 + re^{i\theta}$ on C_1,

$$\int_{C_1} f(z)\,dz \xrightarrow[r \to 0]{} -i\pi a_{-1}.$$

Taking the limit $r \to 0$ in Eq. (8.23) yields

$$P \int_a^b f(x)\,dx = \pi i [\text{residue of } f(z) \text{ at } x_0]$$

$$+ 2\pi i [\text{sum of residues of } f(z) \text{ enclosed by } C] - \int_{C_2} f(z)\,dz. \tag{8.24}$$

Usually the contour is chosen so that the integral over C_2 vanishes.

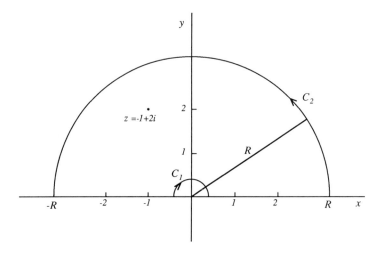

Figure 8.8

Example 8. Calculate the Cauchy principal value of the integral

$$\int_{-\infty}^{\infty} \frac{dx}{x(x^2 + 2x + 5)}. \tag{8.25}$$

Complete the contour off the real axis, as shown in Figure 8.8, and note that the integral along the large semicircle C_2 vanishes in the limit $R \to \infty$. The small semicircle excludes the pole on the real axis at $z = 0$ with residue $\frac{1}{5}$, whereas the complete contour encloses the simple pole at $z = -1 + 2i$, where the residue is $(-2 + i)/20$. Thus, according to Eq. (8.24),

$$P \int_{-\infty}^{\infty} \frac{dx}{x(x^2 + 2x + 5)} = \frac{-\pi}{10}. \tag{8.26}$$

The integrand in Eq. (8.25) is plotted in Figure 8.9. The areas between the curve and the real axis for $x \geq 0$ and $x \leq 0$ separately are infinitely large, but one is positive and the other negative so that there is a cancellation such that the full integral remains finite. From the asymmetry seen in Figure 8.9 it is clear why the Cauchy principal value in Eq. (8.26) is negative.

8.2.3 A BRANCH POINT

Example 9. Now consider the application of the residue theorem when the integrand has a branch point. Let z be a complex number with $0 < \mathrm{Re}(z) < 1$. Then, the integrand of the integral

$$\oint_C \frac{t^{z-1}}{1 + t^2} \, dt$$

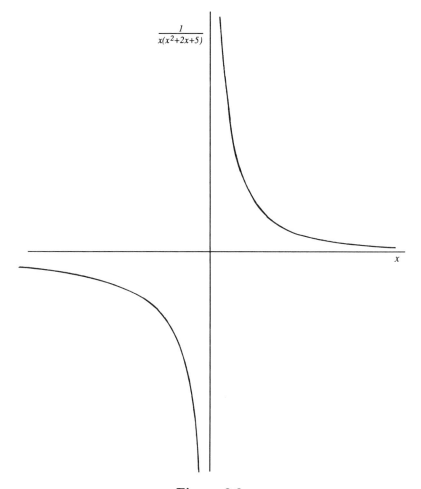

$$\frac{1}{x(x^2+2x+5)}$$

x

Figure 8.9

has a branch point at $t = 0$. Take C to be the contour in Figure 8.10 consisting of two concentric circular arcs of radii r and R connected by straight lines parallel to the real axis.[2]

The region bounded by C excludes the branch point but includes the two simple poles on the imaginary axis at $t = \pm i$. First, apply the residue theorem to obtain

$$\oint_C \frac{t^{z-1}}{1+t^2}\, dt = -2\pi i e^{iz\pi} \cos\left(\frac{z\pi}{2}\right). \qquad (8.27)$$

----

[2] We have made a *branch cut* along the positive real axis.

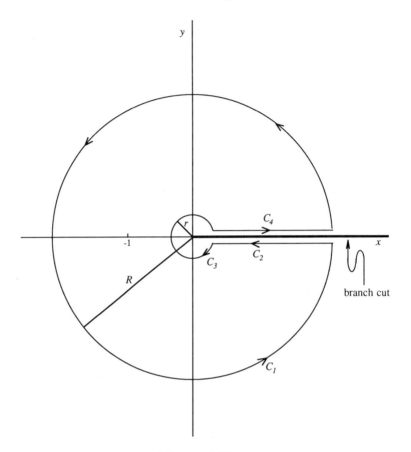

Figure 8.10

Now carry out the integration piecewise over the different parts of the contour to get

$$\oint_C \frac{t^{z-1}}{1+t^2}\, dt = \int_{C_1} \frac{t^{z-1}}{1+t^2}\, dt + \int_{C_2} \frac{t^{z-1}}{1+t^2}\, dt + \int_{C_3} \frac{t^{z-1}}{1+t^2}\, dt + \int_{C_4} \frac{t^{z-1}}{1+t^2}\, dt$$

$$= iR^z \int_0^{2\pi} \frac{e^{iz\theta}}{1+R^2 e^{i2\theta}}\, d\theta + e^{i2\pi z} \int_R^r \frac{x^{z-1}}{1+x^2}\, dx$$

$$+ ir^z \int_{2\pi}^0 \frac{e^{iz\theta}}{1+r^2 e^{i2\theta}}\, d\theta + \int_r^R \frac{x^{z-1}}{1+x^2}\, dx.$$

Next, we take the limits $r \to 0$ and $R \to \infty$. The first integral vanishes, because

$$(R^z/R^2) \xrightarrow[R\to\infty]{} 0 \quad \text{with} \quad 0 < \text{Re}(z) < 1.$$

Similarly, the third integral vanishes because $r^z \xrightarrow[r \to 0]{} 0$. The second and fourth integrals may be combined so that,

$$\oint_C \frac{t^{z-1}}{1+t^2} \, dt = (1 - e^{i2z\pi}) \int_0^\infty \frac{x^{z-1}}{1+x^2} \, dx.$$

With the value of the contour integral given by Eq. (8.27) the definite integral is

$$\int_0^\infty \frac{x^{z-1}}{1+x^2} \, dx = \frac{\pi}{2 \sin\left(\frac{z\pi}{2}\right)}. \tag{8.28}$$

EXERCISES

1. Use the Cauchy residue theorem to obtain the integral representation of $\sin x$,

$$\sin x = \frac{1}{2\pi i} \oint_C \frac{e^{zx}}{z^2 + 1} \, dz$$

where C is any circle about the origin with radius greater than one.[3]

2. Let b and c be real, positive numbers with $b < 2\sqrt{c}$. Use the residue theorem to prove that

$$\int_{-\infty}^\infty \frac{dx}{(x^2 - bx + c)^2} = \frac{4\pi}{[4c - b^2]^{3/2}}.$$

3. Use Jordan's lemma to show that for $\mu > 0$

$$\int_0^\infty \frac{\cos \mu x}{x^2 + 1} \, dx = \frac{\pi}{2} e^{-\mu}.$$

4. Use the method described in Example 4 to evaluate the integral

$$I = \int_0^{2\pi} \frac{\sin^2 \theta}{\mu + \nu \cos \theta} \, d\theta$$

with μ and ν real and $\mu > \nu > 0$. On making the change of variable $z = e^{i\theta}$ it can be seen that the integrand has a double pole at the origin and two simple poles on the real axis. Show that one of the simple poles lies outside the unit circle about the origin and the other one lies inside this circle. Calculate the residues and show that

$$I = \frac{2\pi}{\nu^2} \left[\mu - \sqrt{\mu^2 - \nu^2} \right].$$

---

[3] We could take C to be *any* closed curve that does not cross itself and encloses the two points on the imaginary axis at $z = \pm i$.

5. By contour integration show that

$$\int_0^{2\pi} \frac{d\theta}{a - b\cos\theta} = \frac{2\pi}{\sqrt{a^2 - b^2}}.$$

When compared with the result in Example 4, in view of the different signs in the denominators of the integrands, is this reasonable? Explain.

6. Show by contour integration that if n is a positive integer, then

$$\int_0^{\pi/2} \cos^{2n}\theta\, d\theta = \frac{\pi\left(\frac{1}{2}\right)_n}{2n!}.$$

Some changes of variable will be useful here.

7. Show that the integral[4]

$$I_n = \int_0^{2\pi} e^{\cos\theta} \cos(n\theta - \sin\theta)\, d\theta$$

can be written as

$$I_n = \frac{-i}{2} \oint_C \left[z^{n-1}e^{1/z} + z^{-n-1}e^z\right] dz$$

where n is a nonnegative integer and C is the unit circle centered at the origin. In the first term change the variable of integration to $y = z^{-1} = e^{-i\theta}$ and integrate over $-C$ to obtain

$$I_n = -i \oint_C z^{-n-1}e^z dz.$$

Now use the residue theorem to show that

$$I_n = \frac{2\pi}{n!}.$$

8. Follow the procedure in Example 4 and show that[5]

$$\int_0^{2\pi} e^{\cos\theta} \cos(\sin\theta) \cos\theta\, d\theta = \pi.$$

---

[4]Exercises 7, 8, 10, and 14 are taken from M.R. Spiegel, *Complex Variables*, McGraw-Hill, New York, 1964, and are used with the permission of McGraw-Hill, Inc. This reference contains a wealth of instructive exercises.
[5] *Ibid.*

9. Show that for any closed contour C that encloses the dashed circle in Figure 8.11, we have

$$\oint_C \frac{e^{xt}}{t(t^2+1)} \, dt = 2\pi i (1 - \cos x).$$

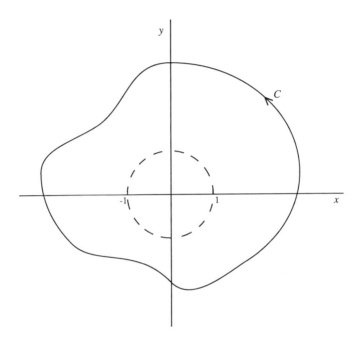

Figure 8.11

10. In the integral

$$\oint_C \frac{e^{zq}}{z^2+p^2} \, dz,$$

p and q are real, positive constants. Let C consist of a line through the point $z = b > 0$ parallel to the imaginary axis and a circular arc of radius R, as in Figure 8.12. Show that taking the limit $R \to \infty$ leads to the result[6]

$$\int_{b-i\infty}^{b+i\infty} \frac{e^{zq}}{z^2+p^2} \, dz = 2\pi i \frac{\sin pq}{p}.$$

Note that in the limit $R \to \infty$ the circular arc is confined to the second and third quadrants where $\cos \theta \le 0$.

---

[6] *Ibid.*

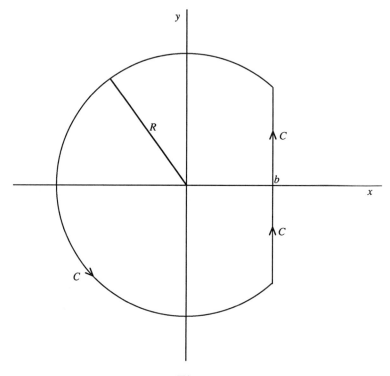

Figure 8.12

11. Define C as the closed contour in Figure 8.13. Then

$$\oint_C \frac{e^{zt}}{\sqrt{z}}\, dz = 0,$$

because the integrand is analytic everywhere on and inside C. This contour avoids the branch point at $z = 0$. More specifically, C consists of six pieces:

$$
\begin{aligned}
&C_1\!: z = b + iy, & -\sqrt{R^2 - b^2} \le y \le \sqrt{R^2 - b^2},\\
&C_2\!: z = Re^{i\theta}, & \cos^{-1}(b/R) < \theta < \pi,\\
&C_3\!: z = ue^{i\pi}, & R \ge u \ge r,\\
&C_4\!: z = re^{i\theta}, & \pi > \theta > -\pi,\\
&C_5\!: z = ue^{-i\pi}, & r \le u \le R,\\
&C_6\!: z = Re^{i\theta}, & -\pi < \theta < -\cos^{-1}(b/R),
\end{aligned}
$$

where b, r, and R are real, positive numbers. Show that in the limits $r \to 0$ and $R \to \infty$ these contributions lead to

$$\int_{b-i\infty}^{b+i\infty} \frac{e^{zt}}{\sqrt{z}}\, dz = 2i\sqrt{\frac{\pi}{t}}.$$

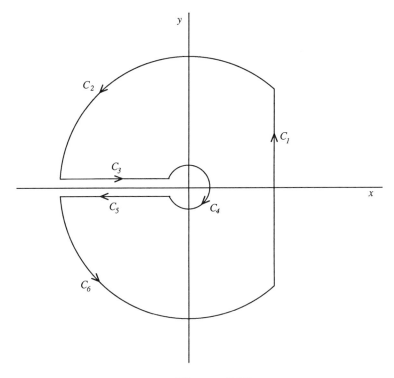

Figure 8.13

The formula in Eq. (1.16) will be useful here.

12. Show that

$$\int_{-\infty}^{\infty} \frac{\sin^2 x}{x^2}\, dx = \pi.$$

Hint: Integrate by parts and take the principal value of the remaining integral.

13. By contour integration show that for $0 < \mathrm{Re}(p) < 1$

$$\int_0^{\infty} \frac{t^{p-1}}{a+t}\, dt = \frac{\pi a^{p-1}}{\sin \pi p}$$

where the point $z = a$ lies anywhere in the complex plane *except* on the negative real axis. With $a = 1$ and $p = \frac{1}{2}$ compare the result here with that in Exercise 1.19.

14. By changing the variable of integration to $x = \tan\phi$ show by integration around the closed contour in Figure 8.14 that[7]

$$\int_0^{\pi/2} \sqrt{\tan\phi}\, d\phi = \frac{\pi}{\sqrt{2}}.$$

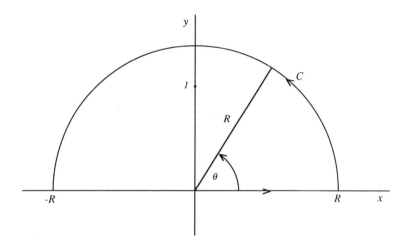

Figure 8.14

15. With a change of variable $x = \tan\phi$ the integral

$$\int_{-\infty}^{\infty} \frac{dx}{(x^2 + 1)^2}$$

may be integrated directly. Show that the result is the same as that obtained in Example 3 by contour integration.

16. Let z be complex. According to the Taylor formula

$$(1 - z)^b = \sum_{k=0}^{n-1} \frac{(-b)_k}{k!} z^k + R_n.$$

Use the residue theorem to show that the remainder term is

$$R_n = \frac{(-b)_n}{n!}(1 - z_1)^{b-n} z^n \quad \text{for} \quad |z_1| < |z| < 1.$$

[7] *Ibid.*

17. In the integral,

$$\oint_C \frac{dz}{\sqrt{z}(1-z)}$$

the integrand has a branch point at the origin and a simple pole on the positive real axis. By choosing an appropriate contour C, evaluate the definite integral

$$\int_0^\infty \frac{dx}{\sqrt{x}(1+x)}.$$

9

Alternate Forms for Special Functions

Authors of different texts on physics and applied mathematics choose different definitions for the special functions. Some obtain series solutions to the differential equation and define the corresponding special function in terms of these series, others prefer to use the generating function as the defining relation, still others elect to start with a Rodrigues formula for the special function, and a few take an integral representation as a definition. It is interesting and instructive to see that these various ways of defining a given special function are all equivalent.

Let us use the mathematical machinery developed in the previous chapters to obtain some of these alternate forms for the special functions we have studied. These expressions are often useful in computing numerical values for the functions. They allow us to extend the domain of validity of the original functions by analytic continuation. They will also provide us with the recursion formulas and orthogonality relations that appear in textbooks on physics and mathematical applications.

9.1 The Gamma Function

From Chapter 1 we have the formula[1]

$$\Gamma(z)\Gamma(1-z) = \int_0^\infty \frac{t^{z-1}}{1+t}\, dt \quad \text{for } 0 < \text{Re}(z) < 1. \tag{9.1}$$

As an illustration of the power of contour integration we now derive a simpler, closed-form expression in terms of elementary functions. Our aim is to find a contour C such that the integral along this path includes the right-hand side of Eq. (9.1) as a part.

The origin is a branch point. Therefore, in the complex plane we must make a cut across which the integrand is discontinuous. For convenience we choose our cut along the negative real axis. This leads us to the contour of Figure 9.1. This contour consists of four pieces. With $u = re^{i\theta}$ we have on the different parts

---

[1] See Eq. (1.20).

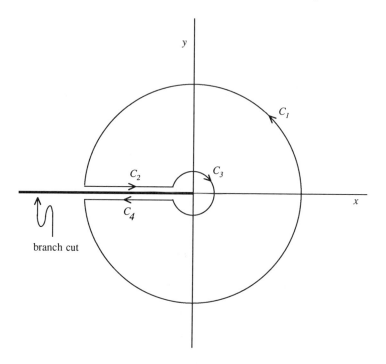

Figure 9.1

$$\begin{aligned}
C_1: & \quad -\pi < \theta < \pi; \ r \to \infty, \\
C_2: & \quad \theta = \pi; \ 0 < r < \infty, \\
C_3: & \quad -\pi < \theta < \pi; \ r \to 0, \\
C_4: & \quad \theta = -\pi; \ 0 < r < \infty.
\end{aligned}$$

Written explicitly in terms of the pieces the integral around C is

$$\oint_C \frac{u^{z-1}}{1-u} \, du = ir^z \int_{C_1} \frac{e^{iz\theta}}{1-re^{i\theta}} \, d\theta + e^{iz\pi} \int_\infty^0 \frac{r^{z-1}}{1+r} \, dr$$
$$+ ir^z \int_{C_3} \frac{e^{iz\theta}}{1-re^{i\theta}} \, d\theta + e^{-iz\pi} \int_0^\infty \frac{r^{z-1}}{1+r} \, dr.$$

Because $0 < z < 1$, the first integral on the right-hand side vanishes as r tends to infinity. The third integral also vanishes as r goes to zero. Combine the remaining two terms to get

$$\oint_C \frac{u^{z-1}}{1-u} \, du = \left(e^{-iz\pi} - e^{i\pi z} \right) \int_0^\infty \frac{r^{z-1}}{1+r} \, dr. \qquad (9.2)$$

The contour encloses a simple pole at $u = +1$. From the residue theorem we find that the integral on the left-hand side is equal to $-2\pi i$. Because the integral on the right-hand side of Eq. (9.2) is the same as the one in Eq. (9.1), it follows that

$$\Gamma(z)\Gamma(1-z) = \frac{\pi}{\sin \pi z}. \tag{9.3}$$

This formula is quite useful later on. With $z = \frac{1}{2}$ we find that $\Gamma(\frac{1}{2}) = \sqrt{\pi}$.

By the principle of analytic continuation Eq. (9.3) holds for every point in the complex plane except at points on the real axis where $z = 0, \pm 1, \pm 2, \ldots$.

9.2 Bessel Functions

In Chapter 4 we obtained as solutions to Bessel's equation the functions $J_\nu(x)$ defined by

$$J_\nu(x) = \frac{1}{\Gamma(\nu+1)} \left(\frac{x}{2}\right)^\nu e^{-ix} \, _1F_1\left(\nu + \tfrac{1}{2}; 2\nu + 1; 2ix\right). \tag{9.4}$$

Another series expression which is often used to *define* $J_\nu(x)$ was given in Eq. (4.19) as

$$J_\nu(x) = \left(\frac{x}{2}\right)^\nu \sum_{k=0}^\infty \frac{(-1)^k}{k! \Gamma(k+\nu+1)} \left(\frac{x}{2}\right)^{2k}$$

$$= \frac{1}{\Gamma(\nu+1)} \left(\frac{x}{2}\right)^\nu \, _0F_1\left(\nu+1; \frac{-x^2}{4}\right). \tag{9.5}$$

To show that this representation of the Bessel function agrees with *our* definition in Eq. (9.4) we have only to show that[2]

$$e^{cz} \, _1F_1(a; 2a; -2cz) = \, _0F_1\left(a + \tfrac{1}{2}; \left[\frac{cz}{2}\right]^2\right). \tag{9.6}$$

First, write the series expansions for the two factors on the left-hand side as a Cauchy product. That is,

$$e^{cz} \, _1F_1(a; 2a; -2cz) = \sum_{n=0}^\infty \sum_{m=0}^n \frac{(cz)^m}{m!} \frac{(a)_{n-m}(-2cz)^{n-m}}{(n-m)!(2a)_{n-m}}$$

$$= \sum_{n=0}^\infty \frac{1}{(2a)_n} \left(\sum_{m=0}^n \frac{(-2)^{n-m}(a)_{n-m}}{(n-m)!} \frac{(2a+n-m)_m}{m!}\right) (cz)^n. \tag{9.7}$$

---

[2]G.N. Watson, *op. cit.*, p. 104.

In the second equality here, we have used the identity

$$(2a)_n = (2a)_{n-m}(2a + n - m)_m.$$

Note that the sum

$$c_p = \sum_{m=0}^{p} \frac{(-2)^{p-m}(a)_{p-m}}{(p-m)!} \frac{(2a + n - m)_m}{m!} \qquad (9.8)$$

looks like the coefficient of the pth term in a Cauchy product expansion,

$$u(t)v(t) = \sum_{p=0}^{\infty} c_p t^p.$$

Therefore, we try to find two analytic functions $u(t)$ and $v(t)$, such that

$$u(t) = \sum_{k=0}^{\infty} \frac{(-2)^k (a)_k}{k!} t^k$$

and

$$v(t) = \sum_{j=0}^{\infty} \frac{(2a + n - j)_j}{j!} t^j.$$

We see that according to Eq. (2.4)

$$u(t) = (1 + 2t)^{-a}.$$

Similarly, for $v(t)$ we have

$$v(t) = (1 + t)^{2a+n-1}.$$

We denote the product of $u(t)$ and $v(t)$ by $w(t)$,

$$w(t) = (1 + 2t)^{-a}(1 + t)^{2a+n-1}.$$

Because u and v are both analytic functions of t, $w(t)$ must also be an analytic function of t, therefore, we can write

$$w(t) = \sum_{p=0}^{\infty} c_p t^p$$

with the coefficients c_p given by Eq. (7.22b), as well as by Eq. (9.8). On comparing Eq. (9.7) and Eq. (9.8) we see that it is the coefficient with $p = n$ that should be evaluated. Therefore, according to Eq. (7.22b),

$$c_n = \frac{1}{2\pi i} \oint_C \frac{(1 + t)^{2a+n-1}(1 + 2t)^{-a}}{t^{n+1}} \, dt.$$

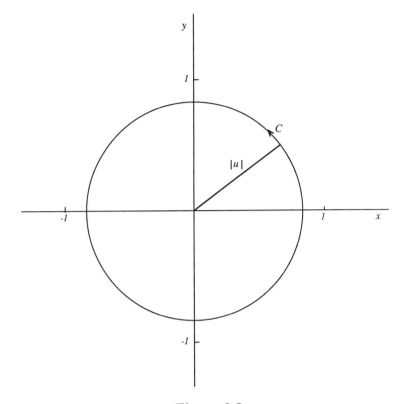

Figure 9.2

The integrand has a pole of order $n+1$ at $t=0$ and is analytic everywhere else inside a circle of radius $|t| < \frac{1}{2}$ centered on the origin. The integral looks formidable in this form. Evaluation is made much simpler with the change of variable $u = t/(1+t)$ which gives

$$c_n = \frac{1}{2\pi i} \oint_C u^{-n-1}(1-u^2)^{-a} du. \qquad (9.9)$$

The integrand in Eq. (9.9) has a pole of order $n+1$ at $u=0$ and no other singularity inside a circle of radius $|u| < 1$ centered at the origin. Therefore, we choose such a circle for our contour C in Eq. (9.9). This path is illustrated in Figure 9.2.

From Eq. (2.4),

$$(1-u^2)^{-a} = \sum_{k=0}^{\infty} \frac{(a)_k}{k!} u^{2k}, \qquad |u| < 1.$$

Make this substitution into Eq. (9.9) and integrate term by term to obtain

$$c_n = \frac{1}{2\pi i} \sum_{k=0}^{\infty} \oint_C \frac{(a)_k}{k!} u^{2k-n-1} \, du. \tag{9.10}$$

For $2k - n - 1 \geq 0$ the integrand is analytic inside C and the integral vanishes. If $2k-n-1 < 0$, then the integrand has a pole of order $-2k+n+1$, in which case the residue theorem can be used to evaluate the integral.

We find that

$$\oint u^{2k-n-1} du = \begin{cases} 2\pi i, & 2k = n, \\ \\ 0, & 2k \neq n. \end{cases}$$

So the only contribution to the sum in Eq. (9.10) comes from the term with $k = n/2$, where n must be even. Thus,

$$c_n = \begin{cases} \dfrac{(a)_{n/2}}{\left(\frac{n}{2}\right)!}, & n = \text{even}, \\ \\ 0, & n = \text{odd}. \end{cases}$$

Use these results for c_n in Eq. (9.8) and then substitute for the corresponding sum in Eq. (9.7) to obtain

$$e^{cz} \, {}_1F_1(a; 2a; -2cz) = \sum_{k=0}^{\infty} \frac{(a)_k}{k!(2a)_{2k}} (cz)^{2k}. \tag{9.11}$$

In Eq. (9.7), $n = 2k$ redefines the summation index, since n must be even. Here the sum is over all integer values of k.

Expand the product in the denominator and reorder the factors to obtain

$$(2a)_{2k} = 2^{2k}(a)_k \left(a + \tfrac{1}{2}\right)_k.$$

With this substitution Eq. (9.11) now reads,

$$e^{cz} \, {}_1F_1(a; 2a; -2cz) = \sum_{k=0}^{\infty} \frac{1}{k!\left(a + \tfrac{1}{2}\right)_k} \left(\frac{cz}{2}\right)^{2k} = {}_0F_1\left(a + \tfrac{1}{2}; \left(\frac{cz}{2}\right)^2\right),$$

$$\tag{9.12}$$

which is the result we set out to obtain in Eq. (9.6). Finally, in Eq. (9.12) we take $a = \nu + \tfrac{1}{2}$ and $c = -i$ and make a substitution with this formula into Eq. (4.17). This brings us to the relation,

$$J_\nu(x) = \frac{1}{\Gamma(\nu+1)} \left(\frac{x}{2}\right)^\nu e^{-ix} \, {}_1F_1\left(\nu + \tfrac{1}{2}; 2\nu + 1; 2ix\right)$$

$$= \frac{1}{\Gamma(\nu+1)} \left(\frac{x}{2}\right)^\nu {}_0F_1\left(\nu + 1; \frac{-x^2}{4}\right),$$

as we wished to show in Eq. (9.5).

The series representation of $J_\nu(x)$ given in Eq. (9.5) is also the starting place for obtaining another interesting and important relation involving Bessel functions. Find a set of functions $g(u, x)$, such that

$$g(u, x) = \sum_{k=-\infty}^{\infty} J_k(x) u^k, \tag{9.13}$$

where each member $g(u, x)$ of the set is labeled by the continuous parameter x and is a function of the variable u.[3]

First, substitute for $J_k(x)$ in Eq. (9.13), the series expansion in Eq. (9.5). This gives

$$g(u, x) = \sum_{m=0}^{\infty} \sum_{k=-\infty}^{\infty} \frac{(-1)^m}{m! \Gamma(k+m+1)} \left(\frac{x}{2}\right)^{k+2m} u^k,$$

where we have reversed the order of summation. Note that $\Gamma(k+m+1)$ diverges for $k < -m$. Therefore, we lose nothing if we write

$$g(u, x) = \sum_{m=0}^{\infty} \sum_{k=-2m}^{\infty} \frac{(-1)^m}{m! \Gamma(k+m+1)} \left(\frac{x}{2}\right)^{k+2m} u^k.$$

Now define a new summation index $n = k + 2m$ and write

$$g(u, x) = \sum_{n=0}^{\infty} \sum_{m=0}^{\infty} \frac{(-1)^m}{m! \Gamma(n-m+1)} \left(\frac{x}{2}\right)^{n} u^{n-2m},$$

where again the order of the sums is reversed. Because the factor $\Gamma(n - m + 1)$ in the denominator diverges for $m > n$, the terms with $m > n$ do not contribute to the sum on m. Therefore, with some rearrangement of the factors in each term

$$g(u, x) = \sum_{n=0}^{\infty} \sum_{m=0}^{n} \left[\left(\frac{(-1)^m}{m!} \left(\frac{x}{2u}\right)^{m}\right) \left(\frac{1}{(n-m)!} \left(\frac{xu}{2}\right)^{n-m}\right) \right].$$

Since n and m are integers, we have written $\Gamma(n - m + 1) = (n - m)!$ This last expression for $g(u, x)$ is manifestly a Cauchy product of two series. Thus,

$$g(u, x) = \sum_{k=0}^{\infty} \frac{1}{k!} \left(\frac{-x}{2u}\right)^{k} \sum_{j=0}^{\infty} \frac{1}{j!} \left(\frac{xu}{2}\right)^{j}$$

$$= e^{\frac{1}{2} x (u - u^{-1})}.$$

---

[3]Because the right hand side has the general form of a Laurent expansion, we should expect that $g(u, x)$ is analytic everywhere outside some deleted neighborhood around $u = 0$.

Finally, from Eq. (9.13)

$$e^{\frac{1}{2}x(u-u^{-1})} = \sum_{k=-\infty}^{\infty} J_k(x)u^k. \tag{9.14}$$

Clearly, Eq. (9.14) is in the form of a Laurent expansion of $\exp\left[\frac{1}{2}x(u - u^{-1})\right]$ about $u = 0$. According to Eqs. (7.22) the expansion coefficients are

$$J_n(x) = \frac{1}{2\pi i} \oint_C u^{-n-1} e^{\frac{1}{2}x(u-u^{-1})} du. \tag{9.15}$$

The contour C encloses the singularity at the origin.[4] This representation of the Bessel function is known as *Schläfli's integral*. From it we are able to derive other useful integral representations of the Bessel functions.

The coefficients generated by the Laurent expansion of $\exp\left[\frac{1}{2}x(u-u^{-1})\right]$ are Bessel functions. For this reason $\exp\left[\frac{1}{2}x(u - u^{-1})\right]$ is called the *generating function* for the Bessel functions. Generating functions for other special functions are discussed in Chapter 11.

Because n is an integer, we can take the contour in Eq. (9.15) to be the unit circle around the origin. In this case $u = e^{i\theta}$ and

$$J_n(x) = \frac{1}{2\pi} \int_0^{2\pi} e^{-in\theta} e^{ix\sin\theta} d\theta$$

$$= \frac{1}{2\pi} \int_0^{\pi} e^{i(x\sin\theta - n\theta)} d\theta + \frac{1}{2\pi} \int_\pi^{2\pi} e^{i(x\sin\theta - n\theta)} d\theta.$$

We have split the range of integration into two parts. In the second part we replace θ by $2\pi - \theta$ to obtain

$$J_n(x) = \frac{1}{\pi} \int_0^{\pi} \cos(x\sin\theta - n\theta) d\theta, \tag{9.16}$$

which is *Bessel's integral*.[5] This integral can be easily evaluated numerically to obtain explicit values for Bessel functions of integer order.[6]

9.3 Legendre Polynomials

Let us show that our definition of the Legendre functions is consistent with other definitions appearing in the literature. In Chapter 5 we defined the

---

[4]The contour must exclude a branch cut if n is not an integer. This case is considered in more detail in Chapter 10.

[5]For the corresponding relation for Bessel functions of noninteger order see Eq. (10.10).

[6]See Figure 4.2.

Legendre polynomials by

$$P_n(x) = {}_2F_1\left(-n, n+1; 1; \frac{1-x}{2}\right).$$

Expand the hypergeometric function and use Eq. (2.4) for $(1-x)^k$ to obtain

$$P_n(x) = \sum_{k=0}^{\infty} \frac{(-n)_k (n+1)_k}{k! k! 2^k} \sum_{m=0}^{\infty} \frac{(k-m+1)_m}{m!} (-1)^m x^m.$$

We reverse the order of the sums and use the fact that $(-n)_k = 0$ for $k > n$ to get

$$P_n(x) = \sum_{m=0}^{\infty} \left(\sum_{k=0}^{n} \frac{(-n)_k (n+1)_k}{k! 2^k} \frac{(k-m+1)_{n-k}}{\Gamma(n-m+1)}\right) \frac{(-1)^m}{m!} x^m.$$

In arriving at this expression we have also used the identity,

$$(k-m+1)_m = k!(k-m+1)_{n-k} / \Gamma(n-m+1).$$

The identities

$$(-n)_k = (-1)^k n!/(n-k)! \quad \text{and} \quad (n+1)_k = (-1)^k(-n-1-k+1)_k,$$

together with the divergence of $\Gamma(n-m+1)$ for $m > n$, allow the double sum to be written as,

$$P_n(x) = \sum_{m=0}^{n} \frac{(-1)^m x^m n!}{m! \Gamma(n-m+1)} \sum_{k=0}^{n} \frac{(-n-1-k+1)_k}{k! 2^k} \frac{(k-m+1)_{n-k}}{(n-k)!}.$$

$$(9.17)$$

The sum

$$c_p = \sum_{r=0}^{p} \frac{(-n-1-r+1)_r}{r! 2^r} \frac{(n-m-(p-r)+1)_{p-r}}{(p-r)!} \qquad (9.18)$$

looks like the coefficient of the pth term in a Cauchy product expansion. Again, look for an analytic function $w(t)$ such that

$$w(t) = \sum_{p=0}^{\infty} c_p t^p.$$

The procedure in Eq. (9.8) *et seq.* leads to the result,

$$w(t) = \left(1 + \frac{t}{2}\right)^{-n-1} (1+t)^{n-m}.$$

This function is analytic near $t = 0$; therefore,

$$c_p = \frac{1}{2\pi i} \oint_C \frac{w(t)}{t^{p+1}} \, dt.$$

The path C encloses the origin, but does not include any singularities of $w(t)$. From a comparison of Eq. (9.17) and Eq. (9.18) it is clear that c_p must be evaluated for $p = n$,

$$c_n = \frac{2^{n+1}}{2\pi i} \oint_C \frac{(1+t)^{n-m}}{(2t+t^2)^{n+1}} \, dt.$$

Now change the variable to $u = 2t + t^2$ to get the integral

$$c_n = \frac{2^n}{2\pi i} \oint_C \frac{(1+u)^{(n-m-1)/2}}{u^{n+1}} \, du$$

$$= \frac{2^n}{2\pi i} \sum_{r=0}^{\infty} \frac{\left(\frac{n-m-1}{2} - r + 1\right)_r}{r!} \oint_C u^{r-n-1} du.$$

Again, we have used the binomial expansion. The closed curve C surrounds the point $u = 0$. In the region bounded by C the integrand is analytic for $r > n$ and has a pole of order $n + 1 - r$ for $r \leq n$. From the formula for residues we find that the residues of all integrals in the sum vanish except for the term $r = n$ in which case the residue is 1. Thus,

$$c_n = \frac{2^n \left(-\frac{1}{2}(n+m-1)\right)_n}{n!}$$

from which the summation

$$P_n(x) = \sum_{m=0}^{n} \frac{(-1)^m x^m 2^n \left(-\frac{1}{2}(n+m-1)\right)_n}{m!(n-m)!}$$

follows. Because $\frac{1}{2}(n+m-1) < n$, we have $\left(-\frac{1}{2}(n+m-1)\right)_n = 0$, unless $n + m - 1$ is equal to an odd integer. This condition is met if a new summation index k is defined by $2k = n - m$. Then

$$P_n(x) = \sum_{k=0}^{[n/2]} \frac{(-1)^{n-2k} x^{n-2k} 2^n}{(n-2k)!(2k)!} \left(k - n + \tfrac{1}{2}\right)_n.$$

Finally, use the identity

$$\frac{\left(k - n + \frac{1}{2}\right)_n}{(2k)!} = 2^{-2n} \frac{(2n-2k)!}{(n-k)!} \frac{(-1)^{k-n}}{k!}$$

to obtain

$$P_n(x) = \sum_{k=0}^{[n/2]} \frac{(-1)^k (2n-2k)!}{2^n k!(n-2k)!(n-k)!} x^{n-2k}. \qquad (9.19)$$

This series is often used as the definition of the Legendre polynomials.[7]
From Eq. (9.19) it is clear that

$$P_n(x) = \sum_{k=0}^{[n/2]} \frac{(-1)^k}{2^n k! (n-k)!} \frac{d^n}{dx^n} x^{2n-2k}.$$

If we carry out the differentiation, we find that

$$\frac{d^n}{dx^n} x^{2n-2k} = (n - 2k + 1)_n x^{n-2k}.$$

Now, $(n - 2k + 1)_n = 0$ for $k \geq \left[\frac{n}{2}\right] + 1$. Therefore, in the sum, terms for $\left[\frac{n}{2}\right] + 1 \leq k \leq n$ can be added without contributing anything new to the sum. Then, by introducing a new summation index $p = n - k$,

$$P_n(x) = \frac{(-1)^n}{2^n n!} \frac{d^n}{dx^n} \sum_{p=0}^{n} \frac{(-1)^p n!}{p!(n-p)!} x^{2p}$$

is obtained. The sum is clearly a binomial expansion; therefore,

$$P_n(x) = \frac{1}{2^n n!} \frac{d^n}{dx^n} (x^2 - 1)^n. \tag{9.20}$$

Also from Eq. (5.22),

$$P_n^m(x) = (1 - x^2)^{\frac{1}{2}|m|} \frac{d^{|m|}}{dx^{|m|}} P_n(x) = \frac{1}{2^n n!} (1 - x^2)^{\frac{1}{2}|m|} \frac{d^{n+|m|}}{dx^{n+|m|}} (x^2 - 1)^n. \tag{9.21}$$

This is *Rodrigues's formula,* which is useful in establishing the orthogonality of the associated Legendre functions and it is sometimes used to *define* the associated Legendre functions.[8]

9.4   Hermite Polynomials

To obtain a Rodrigues expression for the Hermite polynomials multiply Eq. (3.12) by e^{-x^2}. This gives for *even n*

$$e^{-x^2} H_n(x) = \frac{(-1)^{-\frac{1}{2}n} n!}{\left(\frac{n}{2}\right)!} \sum_{m=0}^{\infty} \frac{(-1)^m}{m!} x^{2m} \sum_{k=0}^{\infty} \frac{\left(-\frac{1}{2}n\right)_k}{k! \left(\frac{1}{2}\right)_k} x^{2k}.$$

---

[7] For example, J. Mathews and R. Walker, *Mathematical Methods of Physics,* Benjamin, New York, 1965, pp. 15 and 163. See also Exercise 5.2.

[8] For example, D.J. Griffiths, *Introduction to Electrodynamics,* Prentice-Hall, Englewood Cliffs, 1981, p. 122, and N.N. Lebedev, *op. cit.,* p. 44.

Here we use the series expansions for e^{-x^2} and $_1F_1\left(\frac{-n}{2};\frac{1}{2};x^2\right)$. Note that the terms corresponding to $k > n/2$ contribute nothing, because $\left(-\frac{1}{2}n\right)_k = 0$ for $k > n/2$.

Writing the Cauchy product of the sums yields

$$e^{-x^2}H_n(x) = \frac{(-1)^{-n/2}n!}{\left(\frac{n}{2}\right)!}\sum_{p=0}^{\infty}\left(\sum_{q=0}^{p}\frac{(-1)^{p-q}\left(-\frac{1}{2}n\right)_q}{(p-q)!q!\left(\frac{1}{2}\right)_q}\right)x^{2p}. \qquad (9.22)$$

Using the identity

$$\left(\tfrac{1}{2}\right)_p = (-1)^{p-q}\left(\tfrac{1}{2}\right)_q\left(\tfrac{1}{2}-p\right)_{p-q},$$

we find that

$$\sum_{q=0}^{p}\frac{(-1)^{p-q}\left(-\frac{1}{2}n\right)_q}{(p-q)!q!\left(\frac{1}{2}\right)_q} = \frac{1}{\left(\frac{1}{2}\right)_p}\sum_{q=0}^{p}\frac{\left(-\frac{1}{2}n\right)_q\left(\frac{1}{2}-p\right)_{p-q}}{q!\,(p-q)!} = \frac{\left(\frac{1}{2}-p-\frac{1}{2}n\right)_p}{\left(\frac{1}{2}\right)_p p!}.$$

Here we have invoked Vandermonde's theorem.

With further manipulations of the Pochhammer symbols we obtain

$$\sum_{q=0}^{p}\frac{(-1)^{p-q}\left(-\frac{1}{2}n\right)_q}{(p-q)!q!\left(\frac{1}{2}\right)_q} = \frac{(-1)^p(2p+n)!}{n!\left(\frac{1}{2}n+1\right)_p(2p)!}.$$

This substitution for the sum in Eq. (9.22) yields

$$e^{-x^2}H_n(x) = (-1)^{-\frac{1}{2}n}\sum_{p=0}^{\infty}\frac{(-1)^p(2p+n)!}{\left(p+\frac{1}{2}n\right)!(2p)!}x^{2p}.$$

Now replace the index p by $p - \frac{1}{2}n$ to obtain

$$e^{-x^2}H_n(x) = (-1)^{-\frac{1}{2}n}\sum_{p=\frac{1}{2}n}^{\infty}\frac{(-1)^{p-\frac{1}{2}n}(2p)!}{p!(2p-n)!}x^{2p-n}$$

$$= (-1)^n\sum_{p=0}^{\infty}\frac{(-1)^p(2p)(2p-1)\cdots(2p-n+1)}{p!}x^{2p-n}.$$

All terms vanish for p in the range $0 \le p < \frac{1}{2}n$; therefore, they contribute nothing to the sum. Clearly,

$$e^{-x^2}H_n(x) = (-1)^n\frac{d^n}{dx^n}\sum_{p=0}^{\infty}\frac{(-1)^p}{p!}x^{2p},$$

in which we recognize the sum as e^{-x^2}. Finally, we come to the result,

$$H_n(x) = (-1)^n e^{x^2}\frac{d^n}{dx^n}e^{-x^2}. \qquad (9.23)$$

By multiplying Eq. (3.13) by e^{-x^2} and proceeding in a fashion similar to that above we find that Eq. (9.23) holds for odd n as well as for even n. We use this formula in Chapter 11 to derive the generating function for Hermite polynomials.

9.5 Laguerre Polynomials and Associated Laguerre Polynomials

A Rodrigues formula for the Laguerre polynomials was given in Chapter 6. From Eq. (6.44),

$$L_k(x) = e^x \frac{d^k}{dx^k}(e^{-x}x^k). \tag{9.24}$$

In Chapter 11 we derive a similar expression for the associated Laguerre functions,[9] which, for completeness, is also given here,

$$L_q^p(x) = (-1)^p \frac{\Gamma(q+1)}{(q-p)!} e^x x^{-p} \frac{d^{q-p}}{dx^{q-p}}(e^{-x}x^q). \tag{9.25}$$

The numbers q and p in Eq. (9.25) are not necessarily integers, but the difference $q - p$ must be a nonnegative integer.

EXERCISES

1. Make a change of variable in the integral in Eq. (9.1) to show that for $0 < z < 1$

$$\int_{-\infty}^{\infty} \frac{e^{zx}}{e^x + 1}\, dx = \frac{\pi}{\sin z\pi}.$$

2. Show by contour integration that for $0 < z < 1$

$$\int_0^{\infty} \frac{t^{z-1}}{(1+t)^{n+1}}\, dt = \frac{\pi(-z+1)_n}{n!\sin \pi z}$$

where n is a nonnegative integer. Show also that this result is equivalent to the result given by Eq. (1.19) with $x = z$ and $y = n + 1 - z$.

3. By contour integration show that for $0 < \mu < 1$

$$P \int_0^{\infty} \frac{x^{-\mu}}{x-1}\, dx = \frac{\pi}{\tan \mu\pi},$$

where P denotes the Cauchy principal value of the integral. Use the contour shown in Figure 9.3. The result in Eq. (9.1) will also be useful here.

---

[9] See Eq. (11.19).

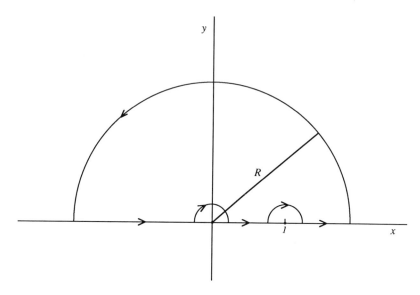

Figure 9.3

4. In Eqs. (8.20) and (8.21) change the variable of integration to $u = x^\nu$ and show that

$$\int_0^\infty \frac{\cos u}{u^p}\, du = \frac{\pi}{2\Gamma(p)\cos\left(\frac{p\pi}{2}\right)} \quad \text{and} \quad \int_0^\infty \frac{\sin u}{u^p}\, du = \frac{\pi}{2\Gamma(p)\sin\left(\frac{p\pi}{2}\right)}$$

with $0 < p < 1$.

5. Expand the product

$$e^{-x}\,{}_1F_1(a;c;x)$$

as a Cauchy product of the series representations of the two functions. By manipulating the Pochhammer symbols and invoking Vandermonde's theorem show that

$${}_1F_1(a;c;x) = e^x\,{}_1F_1(c-a;c;-x).$$

This is Kummer's transformation.[10]

6. Using the formula in Exercise (1.18) show that

$$\int_0^b \frac{dx}{\sqrt{b^3 - x^3}} = \frac{\Gamma\left(\frac{1}{3}\right)\Gamma\left(\frac{1}{6}\right)}{6\sqrt{\pi b}}.$$

---

[10] A. Erdélyi, *op. cit.*, p. 253. For an alternate derivation see Exercise 10.4.

7. Use the generating function for Bessel functions to prove that

$$e^{i\rho\sin\theta} = \sum_{n=-\infty}^{\infty} J_n(\rho)e^{in\theta}.$$

8. Obtain from the series representation of $J_\nu(z)$ the result

$$J_\nu(-z) = e^{i\nu\pi}J_\nu(z).$$

9. Use the definitions of $N_\nu(z)$, $H_\nu^{(1)}(z)$, and $H_\nu^{(2)}(z)$ to show that

$$H_\nu^{(1)}(z) = e^{-i\nu\pi}H_{-\nu}^{(1)}(z)$$

and

$$H_\nu^{(2)}(z) = e^{i\nu\pi}H_{-\nu}^{(2)}(z).$$

10. Show that

$$H_\nu^{(2)}(z) = -e^{i\nu\pi}H_\nu^{(1)}(-z).$$

11. For ν and z both real and nonnegative show that for small values of z

$$J_\nu(z) \approx \frac{1}{\Gamma(\nu+1)}\left(\frac{z}{2}\right)^\nu, \qquad z \to 0.$$

Use the series expansion of $J_\nu(z)$ to establish this result. Show also that for these same values of ν and z we have the approximation

$$N_\nu(z) \approx \frac{-\Gamma(\nu)}{\pi}\left(\frac{2}{z}\right)^\nu, \qquad z \to 0.$$

12. Use Leibniz's theorem to prove that the Laguerre polynomial can also be expressed as

$$L_k(x) = \left(\frac{d}{dx} - 1\right)^k x^k.$$

10

Integral Representations of Special Functions

It is often convenient to identify the various special functions with contour integrals along certain paths in the complex plane. These integrals provide recursion formulas, asymptotic forms, and analytic continuations of the special functions. Also, they are sometimes used as definitions of special functions.[1]

10.1 The Gamma Function

As our first example, we return to the ubiquitous gamma function. Equation (1.13) defined this function as

$$\Gamma(z) = \int_0^\infty t^{z-1} e^{-t}\, dt. \tag{10.1}$$

Now try to find a contour integral that includes this as one piece. The integral

$$\int_C s^{z-1} e^s\, ds$$

suits this purpose. For $0 < \mathrm{Re}(z) < 1$ the integrand is multivalued with a branch point at the origin. Make a branch cut along the negative real axis and choose a contour C that starts below the branch cut, circles the origin in a counterclockwise direction, and then parallels the negative real axis just above the cut as shown in Figure 10.1.

Set $s = u e^{i\theta}$ and evaluate the integrals for the three pieces separately,

$$
\int_C s^{z-1} e^s\, ds = \lim_{r \to 0} \Bigg[\int_\infty^r e^{-u} (u e^{-i\pi})^{z-1} e^{-i\pi}\, du
$$
$$
+ ir^z \int_{\mathrm{arc}} e^{r e^{i\theta}} (e^{i\theta})^{z-1} e^{i\theta}\, d\theta
$$
$$
+ \int_r^\infty e^{-u} (u e^{i\pi})^{z-1} e^{i\pi}\, du \Bigg].
$$

---

[1] For example, E. Merzbacher, *op. cit.*, p. 194, and A.L. Fetter and J.D. Walecka, *op. cit.*, pp. 532 and 541.

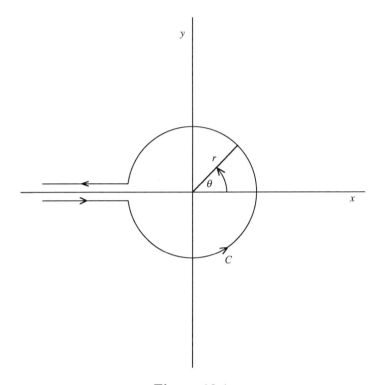

Figure 10.1

For $\mathrm{Re}(z) > 0$ the second integral on the right vanishes as r shrinks to zero. After taking the limit, we are left with

$$\int_C s^{z-1}e^s\,ds = \left(e^{i\pi z} - e^{-i\pi z}\right)\int_0^\infty u^{z-1}e^{-u}\,du.$$

On comparing this result with Eq. (10.1) we see that

$$\int_C s^{z-1}e^s\,ds = 2i\sin\pi z\,\Gamma(z).$$

Thus, the integral representation is

$$\Gamma(z) = \frac{1}{2i\sin\pi z}\int_C s^{z-1}e^s\,ds, \tag{10.2}$$

where C is the contour in Figure 10.1. This function is analytic for all z not equal to an integer. Therefore, Eq. (10.2) provides an analytic continuation of $\Gamma(z)$ to the left-hand half plane.

By rearranging Eq. (9.3), we get

$$\frac{1}{\Gamma(z)} = \frac{\sin \pi z}{\pi}\Gamma(1 - z).$$

Now use Eq. (10.2) to evaluate $\Gamma(1 - z)$. Thereby, another integral representation for $\Gamma(z)$ is

$$\frac{1}{\Gamma(z)} = \frac{1}{2\pi i} \int_C t^{-z} e^t \, dt, \tag{10.3}$$

where again C is the contour in Figure 10.1. This will soon be useful.

Let us make a more detailed examination of the analyticity of the gamma function.[2] Clearly, the integrand in Eq. (10.1) has no singularities for $\text{Re}(z) \geq 1$, but for $\text{Re}(z) < 1$ it is singular at the lower limit, $t = 0$. By writing the integral as two pieces,

$$\Gamma(z) = \int_0^1 t^{z-1} e^{-t} dt + \int_1^\infty t^{z-1} e^{-t} \, dt, \tag{10.4}$$

we see that the second term is analytic everywhere because its derivative exists and the range of integration does not include any singularities of the integrand.

Now expand e^{-t} in the first term of Eq. (10.4) and reverse the order of summation and integration. This gives,

$$\int_0^1 t^{z-1} e^{-t} \, dt = \sum_{k=0}^\infty \frac{(-1)^k}{k!} \int_0^1 t^{z+k-1} \, dt$$

$$= \sum_{k=0}^\infty \frac{(-1)^k}{k!} \frac{1}{z+k}.$$

Thus,

$$\Gamma(z) = \sum_{k=0}^\infty \frac{(-1)^k}{k!} \frac{1}{z+k} + \int_1^\infty t^{z-1} e^{-t} \, dt, \tag{10.5}$$

from which it is clear that the singularities of $\Gamma(z)$ are simple poles on the real axis at $z = 0, -1, -2, -3, \ldots$, and so on.[3]

10.2   Bessel Functions

The gamma function can be used to obtain a contour integral representation of the Bessel function of order ν. Let us seek a function $f(x, t)$ such that

$$J_\nu(x) = \oint_C f(x, t) \, dt = 2\pi i[\text{sum of residues}]. \tag{10.6}$$

---

[2]N.N. Lebedev, *op. cit.*, p. 1.
[3]See also Exercise 1.10.

A comparison of Eq. (10.6) with Eq. (9.5) shows that

$$\text{sum of residues} = \frac{1}{2\pi i}\sum_{k=0}^{\infty}\frac{(-1)^k}{k!\,\Gamma(\nu+k+1)}\left(\frac{x}{2}\right)^{2k+\nu}$$

$$= \sum_{k=0}^{\infty}a_{-1}(t_k). \tag{10.7}$$

Therefore, we look for a function that has an infinite number of discrete poles. One such function is $\Gamma(-t)$ with simple poles at $t = 0, 1, 2, \ldots$ as seen in Eq. (10.5). For t near some positive integer n it is only the nth term in the sum of Eq. (10.5) that is singular. In this case

$$\Gamma(-t) = \frac{(-1)^n}{n!}\frac{1}{-t+n} + \Lambda(t) \quad \text{for } n-1 < \text{Re}(t) < n+1,$$

where $\Lambda(t)$ is analytic in this region. Thus,

$$\lim_{t\to n}\left[(t-n)\Gamma(-t)\right] = \frac{(-1)^{n+1}}{n!}.$$

From Eq. (10.7) it follows that

$$a_{-1}(t_n) = \lim_{t\to n}\left[(t-n)f(x,t)\right]$$

$$= \lim_{t\to n}\left[\frac{-1}{2\pi i}\frac{(t-n)\Gamma(-t)}{\Gamma(\nu+t+1)}\left(\frac{x}{2}\right)^{2t+\nu}\right].$$

Clearly,

$$f(x,t) = \frac{-1}{2\pi i}\frac{\Gamma(-t)}{\Gamma(\nu+t+1)}\left(\frac{x}{2}\right)^{2t+\nu},$$

which has poles on the real axis at $t = 0, 1, 2, \ldots$. Thus,

$$J_\nu(x) = \frac{-1}{2\pi i}\oint_C \frac{\Gamma(-t)}{\Gamma(\nu+t+1)}\left(\frac{x}{2}\right)^{2t+\nu}dt, \tag{10.8}$$

where the contour C shown in Figure 10.2 is closed at $t = +\infty$. Distorting the contour to that in Figure 10.3 yields Barnes's representation,[4]

$$J_\nu(x) = \frac{1}{2\pi i}\int_{-i\infty}^{i\infty}\frac{\Gamma(-t)}{\Gamma(\nu+t+1)}\left(\frac{x}{2}\right)^{2t+\nu}dt,$$

which is valid for $\text{Re}(\nu) > 0$.

In Section 9.2, we obtained Schläfli's integral representation for Bessel functions of integer order. Now let us show that Eq. (9.15) is also applicable

---

[4]See G.N. Watson, *op. cit.*, p. 192.

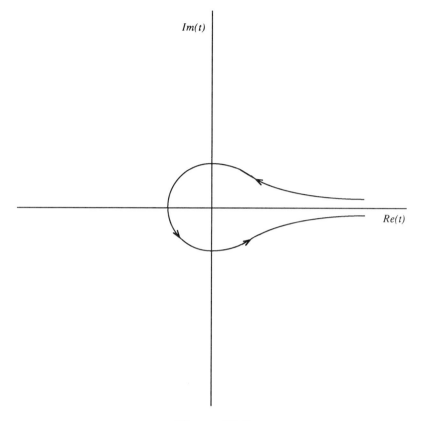

Figure 10.2

to Bessel functions of noninteger order. We start with the series expansion in Eq. (9.5),

$$J_\nu(x) = \left(\frac{x}{2}\right)^\nu \sum_{k=0}^{\infty} \frac{(-1)^k}{k!\Gamma(k+\nu+1)} \left(\frac{x}{2}\right)^{2k},$$

and substitute for the gamma function from Eq. (10.3). This yields

$$J_\nu(x) = \frac{1}{2\pi i} \left(\frac{x}{2}\right)^\nu \int_C t^{-\nu-1} e^t \left[\sum_{k=0}^{\infty} \frac{(-1)^k}{k!} \left(\frac{x^2}{4t}\right)^k\right] dt$$

where the order of summation and integration has been reversed. The sum in the brackets is the Taylor series for $\exp(-x^2/4t)$. Thus,

$$J_\nu(x) = \frac{1}{2\pi i} \left(\frac{x}{2}\right)^\nu \int_C t^{-\nu-1} e^{(t-x^2/4t)} \, dt. \qquad (10.9)$$

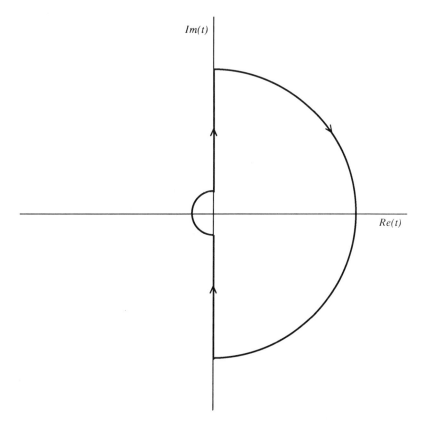

Figure 10.3

This relation holds for noninteger values of ν as well as for integer values. It may be used to derive the basic recursion relations for Bessel functions.[5] It also can be used to show that Schläfli's integral (Eq. 9.15) is valid for Bessel functions of noninteger order.

Changing the variable to $u = 2t/x$ in Eq. (10.9) yields

$$J_\nu(x) = \frac{1}{2\pi i} \int_C u^{-\nu-1} e^{\frac{1}{2}x(u-u^{-1})} \, du.$$

The contour C consists of paths parallel to the branch cut along the negative real axis connected by an arc of the unit circle as shown in Figure 10.1.

---

[5] For example, see Exercise 10.2.

From this last expression a generalization of Bessel's integral can be derived. Integration over the three parts of this contour separately yields

$$J_\nu(x) = \frac{-1}{2\pi i}\left(e^{i\nu\pi} - e^{-i\nu\pi}\right)\int_1^\infty r^{-\nu-1}e^{-\frac{1}{2}x(r-r^{-1})}\,dr$$
$$+ \frac{1}{2\pi}\int_{-\pi}^{\pi} e^{i(x\sin\theta - \nu\theta)}\,d\theta.$$

In the first integral the variable of integration is redefined according to $r = e^t$. The second integral can be written as two parts: one for $-\pi \le \theta \le 0$ and the other for $0 \le \theta \le \pi$. With these modifications

$$J_\nu(x) = \frac{1}{\pi}\int_0^\pi \cos(x\sin\theta - \nu\theta)\,d\theta - \frac{\sin\nu\pi}{\pi}\int_0^\infty e^{-(x\sinh t + \nu t)}\,dt. \quad (10.10)$$

This is a generalization of Bessel's integral (Eq. 9.16) to Bessel functions of noninteger order.

Now consider Poisson's integral again. We recognize that the integrand in Eq. (4.29) is an even function of θ, which allows symmetrical integration about the origin from $\theta = -\pi/2$ to $\theta = \pi/2$. Now replace θ by $\pi/2 - \theta$ and obtain the formula

$$J_\nu(x) = \frac{1}{\sqrt{\pi}\,\Gamma\left(\nu + \frac{1}{2}\right)}\left(\frac{x}{2}\right)^\nu \int_0^\pi \cos(x\cos\theta)\sin^{2\nu}\theta\,d\theta.$$

Slightly different forms of Poisson's integral may be obtained from the variable change $t = \sin\theta$ in Eq. (4.29),

$$J_\nu(x) = \frac{2}{\sqrt{\pi}\,\Gamma\left(\nu + \frac{1}{2}\right)}\left(\frac{x}{2}\right)^\nu \int_0^1 \cos(xt)(1-t^2)^{\nu-\frac{1}{2}}\,dt. \quad (10.11)$$

Since the integrand is an even function of t, we can also write

$$J_\nu(x) = \frac{1}{\sqrt{\pi}\,\Gamma\left(\nu + \frac{1}{2}\right)}\left(\frac{x}{2}\right)^\nu \int_{-1}^1 \cos(xt)(1-t^2)^{\nu-\frac{1}{2}}\,dt. \quad (10.12)$$

If we use $\cos(xt) = \frac{1}{2}\left(e^{ixt} + e^{-ixt}\right)$ and replace t by $-t$ in the second integral, Eq. (10.12) becomes

$$J_\nu(x) = \frac{1}{\sqrt{\pi}\,\Gamma\left(\nu + \frac{1}{2}\right)}\left(\frac{x}{2}\right)^\nu \int_{-1}^1 e^{ixt}(1-t^2)^{\nu-\frac{1}{2}}\,dt. \quad (10.13)$$

Again replace t by $-t$ to get

$$J_\nu(x) = \frac{1}{\sqrt{\pi}\,\Gamma\left(\nu + \frac{1}{2}\right)}\left(\frac{x}{2}\right)^\nu \int_{-1}^1 e^{-ixt}(1-t^2)^{\nu-\frac{1}{2}}\,dt. \quad (10.14)$$

10.3   Spherical Bessel Functions

The spherical Bessel functions have extremely important applications in physics. We had a glimpse of this in Chapter 6 when analyzing the free particle and a particle whose potential energy makes a sudden change from one constant value to another. In practice it is often necessary to impose physical boundary conditions for such particles at large distances from the origin. Therefore, it is useful to derive finite series expansions for these functions and to examine their behavior for large values of the argument.

With $\nu = \frac{1}{2}$ Eq. (10.11) yields

$$J_{\frac{1}{2}}(x) = \sqrt{\frac{2x}{\pi} \frac{\sin x}{x}}.$$

Thus, from the definition of the spherical Bessel functions (Eq. 6.9a),

$$j_0(x) = \frac{\sin x}{x}. \tag{10.15}$$

For $l = 0$ Eq. (6.22) gives

$$j_1(x) = \frac{\sin x}{x^2} - \frac{\cos x}{x}. \tag{10.16}$$

From Eqs. (6.9) and the recursion formula (Eq. 4.21) with $\nu = \frac{1}{2}$, we get

$$n_0(x) = -\frac{1}{x} j_0(x) + j_1(x).$$

Substitutions from Eqs. (10.15) and (10.16) then lead to the result

$$n_0(x) = -\frac{\cos x}{x}. \tag{10.17}$$

Equation (6.27) shows that all of the functions $j_l(x)$ and $n_l(x)$ may be expressed in terms of trigonometric functions. Specifically,

$$j_l(x) = (-1)^l x^l \left(\frac{1}{x} \frac{d}{dx}\right)^l \left[\frac{\sin x}{x}\right], \tag{10.18}$$

$$n_l(x) = -(-1)^l x^l \left(\frac{1}{x} \frac{d}{dx}\right)^l \left[\frac{\cos x}{x}\right]. \tag{10.19}$$

With these formulas the spherical Hankel functions defined in Eqs. (6.14) may be written as

$$h_l^{(1)}(x) = -i(-1)^l x^l \left(\frac{1}{x} \frac{d}{dx}\right)^l \left[\frac{e^{ix}}{x}\right] \tag{10.20}$$

and

$$h_l^{(2)}(x) = i(-1)^l x^l \left(\frac{1}{x}\frac{d}{dx}\right)^l \left[\frac{e^{-ix}}{x}\right]. \tag{10.21}$$

With successive applications of the differential operator $\frac{1}{x}\frac{d}{dx}$ we get

$$\left(\frac{1}{x}\frac{d}{dx}\right)^2 = \frac{1}{x^2}\frac{d^2}{dx^2} - \frac{1}{x^3}\frac{d}{dx},$$

$$\left(\frac{1}{x}\frac{d}{dx}\right)^3 = \frac{1}{x^3}\frac{d^3}{dx^3} - \frac{3}{x^4}\frac{d^2}{dx^2} + \frac{3}{x^5}\frac{d}{dx},$$

and, in general,

$$\left(\frac{1}{x}\frac{d}{dx}\right)^l = \sum_{k=0}^{l-1} c_k^l \frac{1}{x^{l+k}}\frac{d^{l-k}}{dx^{l-k}}. \tag{10.22}$$

Now apply Eq. (10.22) to both sides of the identity,

$$\left(\frac{1}{x}\frac{d}{dx}\right)^l = \frac{1}{x}\frac{d}{dx}\left(\frac{1}{x}\frac{d}{dx}\right)^{l-1}$$

to get

$$\sum_{k=0}^{l-1} c_k^l \frac{1}{x^{l+k}}\frac{d^{l-k}}{dx^{l-k}} = \sum_{k=0}^{l-2} c_k^{l-1} \frac{1}{x^{l+k}}\frac{d^{l-k}}{dx^{l-k}}$$

$$- \sum_{k=0}^{l-2} c_k^{l-1} \frac{(l+k-1)}{x^{l+k+1}}\frac{d^{l-k-1}}{dx^{l-k-1}}.$$

After combining terms of the same order in the differential operator this equation becomes

$$(c_0^l - c_0^{l-1})\frac{1}{x}\frac{d^l}{dx^l} + (c_{l-1}^l + (2l-3)c_{l-2}^{l-1})\frac{1}{x^{2l-1}}\frac{d}{dx}$$

$$+ \sum_{k=1}^{l-2}[c_k^l - c_k^{l-1} + (l+k-2)c_{k-1}^{l-1}]\frac{1}{x^{l+k}}\frac{d^{l-k}}{dx^{l-k}} = 0.$$

The differential operators are independent of each other. Thus, the coefficient of each operator is zero. This leads to the following recursion formulas for the coefficients c_k,

$$c_0^l = c_0^{l-1} = c_0^1 = 1,$$
$$c_k^l = c_k^{l-1} - (l+k-2)c_{k-1}^{l-1}, \qquad 1 \le k \le l-2,$$
$$c_{l-1}^l = -(2l-3)c_{l-2}^{l-1}, \qquad l \ge 2,$$
$$c_k^l = 0, \qquad k \ge l.$$

If the expansion given in Eq. (10.22) is substituted for the operator in Eq. (10.20), then for $l > 0$

$$h_l^{(1)}(x) = -i(-1)^l x^l \sum_{k=0}^{l-1} c_k^l \frac{1}{x^{l+k}} \frac{d^{l-k}}{dx^{l-k}} \left(\frac{e^{ix}}{x} \right). \qquad (10.23)$$

From Leibniz's theorem we get

$$\frac{d^{l-k}}{dx^{l-k}} \left(\frac{e^{ix}}{x} \right) = \sum_{m=0}^{l-k} \frac{(l-k-m+1)_m}{m!} \frac{d^m}{dx^m} \left(\frac{1}{x} \right) \frac{d^{l-k-m}}{dx^{l-k-m}} e^{ix}$$

$$= i^{l-k} e^{ix} \sum_{m=0}^{l-k} i^{-m}(-1)^m (l-k-m+1)_m \frac{1}{x^{m+1}}.$$

Finally, insert this result into Eq. (10.23) to obtain

$$h_l^{(1)}(x) = -i \frac{e^{i(x-l\pi/2)}}{x} \sum_{k=0}^{l-1} \sum_{m=0}^{l-k} (l-k-m+1)_m c_k^l i^{m-k} \frac{1}{x^{k+m}}, \quad l > 0. \qquad (10.24)$$

Similarly, for the spherical Hankel function of the second kind

$$h_l^{(2)}(x) = i \frac{e^{-i(x-l\pi/2)}}{x} \sum_{k=0}^{l-1} \sum_{m=0}^{l-k} c_k^l (l-k-m+1)_m i^{k-m} \frac{1}{x^{k+m}}, \quad l > 0. \qquad (10.25)$$

Clearly, the sums in Eqs. (10.24) and (10.25) are expansions in powers of $1/x$. These expansions are finite with $l(l+3)/2$ terms and they are exact. For large x the dominant term in the double sum corresponds to $k = m = 0$. Thus, for large x we have the asymptotic forms,

$$h_l^{(1)}(x) \xrightarrow[x\to\infty]{} -i \frac{e^{i(x-l\pi/2)}}{x} \qquad (10.26)$$

and

$$h_l^{(2)}(x) \xrightarrow[x\to\infty]{} i \frac{e^{-i(x-l\pi/2)}}{x}, \qquad (10.27)$$

which are valid for $l \geq 0$. From these expressions and Eqs. (6.14) it follows that

$$j_l(x) \xrightarrow[x\to\infty]{} \frac{\sin(x-l\pi/2)}{x} \qquad (10.28)$$

and

$$n_l(x) \xrightarrow[x\to\infty]{} -\frac{\cos(x-l\pi/2)}{x}. \qquad (10.29)$$

10.4   Legendre Polynomials

A contour integral representation of the Legendre polynomials is easily obtained from Rodrigues's formula of Chapter 9. According to Eq. (7.14)

$$\frac{d^n}{dz^n} f(z) = \frac{n!}{2\pi i} \oint_C \frac{f(t)}{(t-z)^{n+1}} dt,$$

provided $f(t)$ has no singularities inside the closed contour C. A comparison with Rodrigues's formula (Eq. 9.20) suggests that we take

$$f(z) = \frac{1}{2^n n!} (z^2 - 1)^n.$$

This gives

$$\frac{1}{2^n n!} \frac{d^n}{dz^n} (z^2 - 1)^n = \frac{1}{2^n} \frac{1}{2\pi i} \oint_C \frac{(t^2 - 1)^n}{(t-z)^{n+1}} dt,$$

where C encloses the point at $t = z$.

Because the left-hand side is $P_n(z)$, we have the integral formula

$$P_n(z) = \frac{1}{2^n} \frac{1}{2\pi i} \oint_C \frac{(t^2 - 1)^n}{(t-z)^{n+1}} dt, \qquad (10.30)$$

which is *Schläfli's integral* for the $P_n(z)$. This integral is especially useful in obtaining recursion formulas for the Legendre functions as we now show. On taking the derivative of Eq. (10.30) we get

$$P'_n(z) = (n+1) \frac{1}{2\pi i} \frac{1}{2^n} \oint_C \frac{(t^2 - 1)^n}{(t-z)^{n+2}} dt. \qquad (10.31)$$

From the identity

$$P_n(z) = \frac{1}{2\pi i} \frac{1}{2^n} \left[\oint_C \frac{t(t^2 - 1)^n}{(t-z)^{n+2}} dt - z \oint_C \frac{(t^2 - 1)^n}{(t-z)^{n+2}} dt \right]$$

we see that Eq. (10.31) may be written

$$P'_n(z) = \frac{n+1}{2^n z} \frac{1}{2\pi i} \oint_C \frac{t(t^2 - 1)^n}{(t-z)^{n+2}} dt - \frac{n+1}{z} P_n(z). \qquad (10.32)$$

Now,

$$\oint_C \frac{d}{dt} \left[\frac{(t^2 - 1)^{n+1}}{(t-z)^{n+2}} \right] dt = 2(n+1) \oint_C \frac{t(t^2 - 1)^n}{(t-z)^{n+2}} dt$$

$$- (n+2) \oint_C \frac{(t^2 - 1)^{n+1}}{(t-z)^{n+3}} dt, \qquad (10.33)$$

where the differentiation has been carried out in the integrand explicitly. The left-hand side vanishes with the exact differential integrated around a closed loop. Also by Eq. (10.31) we recognize $P'_{n+1}(z)$ in the second term on the right-hand side of Eq. (10.33). Thus,

$$(n+1) \oint_C \frac{t(t^2-1)^n}{(t-z)^{n+2}}\, dt = \frac{n+2}{2} \oint_C \frac{(t^2-1)^{n+1}}{(t-z)^{n+3}}\, dt = 2\pi i 2^n P'_{n+1}(z)$$

and Eq. (10.32) reduces to the recursion formula

$$zP'_n(z) + (n+1)P_n(z) - P'_{n+1}(z) = 0. \qquad (10.34)$$

Later, we shall also use Schläfli's integral to derive the generating function for the Legendre polynomials.[6]

Another integral representation of $P_n(z)$ may be obtained in the following way. Choose for the contour C in Eq. (10.30) a circle of radius $\left|\sqrt{z^2-1}\right|$ centered on the point z, as in Figure 10.4. Then,

$$t = z + \left|\sqrt{z^2-1}\right| e^{i\theta}.$$

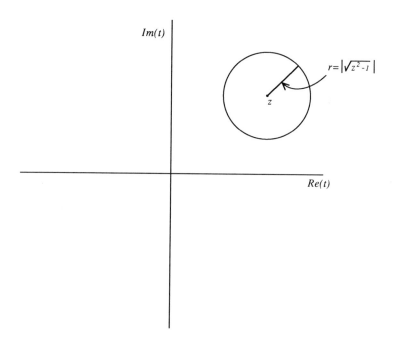

Figure 10.4

---

[6]Section 11.3.

Since

$$\sqrt{z^2 - 1} = \left|\sqrt{z^2 - 1}\right|e^{i\alpha},$$

where α is a constant for fixed z, we can write

$$t = z + \sqrt{z^2 - 1}\, e^{i\phi}.$$

Here $\phi = \theta - \alpha$. With this change of variable Eq. (10.30) can be written

$$P_n(z) = \frac{1}{2\pi} \int_{-\pi}^{\pi} \left(z + \sqrt{z^2 - 1}\cos\phi\right)^n d\phi.$$

Replace ϕ by $-\phi$ over the range $-\pi \leq \phi \leq 0$ to obtain the integral representation[7]

$$P_n(z) = \frac{1}{\pi} \int_0^{\pi} \left(z + \sqrt{z^2 - 1}\cos\phi\right)^n d\phi \tag{10.35}$$

for the Legendre polynomials.

10.5 Laguerre Functions

An integral representation of the associated Laguerre function $L_q^p(x)$ may be derived from the Rodrigues formula of Eq. (9.25). With $\sqrt{x} = s/2r$ and $\sqrt{u} = ry$ in Eq. (4.20) we get

$$x^\nu e^{-x} = \int_0^\infty (xu)^{\frac{1}{2}\nu} J_\nu(2\sqrt{xu})e^{-u}du.$$

Differentiate this expression m times with respect to x. The formula in Exercise 4.5 (with $xu = t$) is useful here to carry out the differentiation under the integral sign:

$$\frac{d^m}{dx^m}[e^{-x}x^\nu] = \int_0^\infty u^m \left[(xu)^{\frac{1}{2}(\nu - m)} J_{\nu - m}(2\sqrt{xu})\right] e^{-u}du.$$

After setting $m = q - p$ and $\nu = q$ here, we see that Eq. (9.25) may be written as

$$L_q^p(x) = (-1)^p \frac{\Gamma(q+1)}{(q-p)!} e^x x^{-\frac{1}{2}p} \int_0^\infty u^{q - \frac{1}{2}p} J_p(2\sqrt{xu})e^{-u}du, \tag{10.36}$$

where q and p are not necessarily integers, but the difference $q - p$ must be a nonnegative integer.

Contour integral representations of the associated Laguerre polynomials are obtained in Eqs. (11.16) and (11.17).

---

[7]This is known as *Laplace's integral representation* for $P_n(z)$. For a more general expression see A. Erdélyi, *op. cit.*, vol. 1, p. 155.

10.6 Hermite Polynomials

For completeness we give here some integral representations of Hermite polynomials, which are derived later. The Hermite polynomial of order n may be represented as a contour integral (see Eq. 11.5).

$$H_n(x) = \frac{n!}{2\pi i} \oint_C \frac{e^{2xt-t^2}}{t^{n+1}} \, dt,$$

where C encloses the origin. The basic recursion relation (Eq. 11.8) follows from this formula. In Chapter 12 (Eq. 12.26) we derive the formula

$$H_n(x) = \frac{i^n}{2\sqrt{\pi}} \int_{-\infty}^{\infty} t^n e^{-\frac{1}{4}(t+2ix)^2} \, dt.$$

10.7 The Hypergeometric Function

An integral representation for the hypergeometric function itself is obtained directly from the definition of Eq. (2.32),

$$_2F_1(a, b; c; z) = \sum_{k=0}^{\infty} \frac{(a)_k (b)_k}{k!(c)_k} z^k. \tag{10.37}$$

If c is not zero or a negative integer, this series converges absolutely for $|z| < 1$. It diverges for $|z| > 1$. For $|z| = 1$ it converges absolutely if $\mathrm{Re}(c - a - b) > 0.$[8]

From the property of the gamma function[9]

$$(s)_k \Gamma(s) = \Gamma(s + k) \tag{10.38}$$

and from Eq. (1.22),

$$\frac{(b)_k}{(c)_k} = \frac{\Gamma(c)}{\Gamma(b)\Gamma(c - b)} \int_0^1 t^{b+k-1}(1 - t)^{c-b-1} \, dt.$$

If this expression is substituted into Eq. (10.37) and the order of summation and integration is changed,

$$_2F_1(a, b; c; z) = \frac{\Gamma(c)}{\Gamma(b)\Gamma(c - b)} \int_0^1 t^{b-1}(1 - t)^{c-b-1} \left[\sum_{k=0}^{\infty} \frac{(a)_k}{k!}(zt)^k \right] dt.$$

---

[8] A. Erdélyi, *op. cit.*, vol. 1, p. 57.
[9] See Exercise 2.5.

The sum on the right is the Taylor expansion of $(1 - zt)^{-a}$. Thus, for $\text{Re}(c) > \text{Re}(b) > 0$ and $|z| < 1$

$$_2F_1(a, b; c; z) = \frac{\Gamma(c)}{\Gamma(b)\Gamma(c - b)} \int_0^1 t^{b-1}(1 - t)^{c-b-1}(1 - zt)^{-a}dt. \quad (10.39)$$

By analytic continuation this integral can be used to define the hypergeometric function for $|z| > 1$.

By a similar technique an integral representation for the confluent hypergeometric function[10] for $\text{Re}(a) > \text{Re}(c) > 0$ is found to be

$$_1F_1(a; c; z) = \frac{\Gamma(c)}{\Gamma(a)\Gamma(c - a)} \int_0^1 e^{tz}t^{a-1}(1 - t)^{c-a-1}dt. \quad (10.40)$$

10.8 Asymptotic Expansions

In quantum mechanics the wave function that describes the behavior of a particle moving in some force field should be well behaved everywhere, including points far from the source of the field. To find the correct form for the wave function at large distances, it is convenient to have asymptotic expansions for such functions valid for large values of the argument.

According to the definition given by Poincaré, the series

$$\sum_{k=0}^{\infty} a_k z^{-k}$$

is the asymptotic series for the function $f(z)$ provided that

$$\lim_{|z| \to \infty} z^n \left[f(z) - \sum_{k=0}^n a_k z^{-k} \right] = 0 \quad \text{for } n > 0.$$

This means that for a given n, if $|z|$ is large enough, the partial sum approximates $f(z)$.[11]

As an example let us look for the asymptotic behavior of the confluent hypergeometric function $_1F_1(a; c; z)$ for large values of $|z|$. For our purposes the most useful situations occur when z is real or purely imaginary. Therefore, we examine four cases:[12] $z = \pm|z|$ and $z = \pm i|z|$.

For $z = |z|$ we start with the integral representation of $_1F_1(a; c; z)$ in Eq. (10.40). Extend the range of integration to negative values of t and then

---

[10]See Exercise 10.3.

[11]For a thorough discussion of asymptotic expansions see E.T. Whittaker and G.N. Watson, *op. cit.*, p. 150.

[12]For complex z see P.M. Morse and H. Feshbach, *op. cit.*, p. 607.

subtract the added piece to obtain

$$_1F_1(a; c; z) = \frac{\Gamma(c)}{\Gamma(a)\Gamma(c-a)} \left[\int_{-\infty}^{1} e^{zt} t^{a-1} (1-t)^{c-a-1} dt \right.$$

$$\left. - \int_{-\infty}^{0} e^{zt} t^{a-1} (1-t)^{c-a-1} dt \right].$$

Change the variable to $t = 1 - u/z$ in the first integral and to $t = -u/z$ in the second integral, so that

$$_1F_1(a; c; z) = \frac{\Gamma(c)}{\Gamma(a)\Gamma(c-a)} \left[\frac{e^z}{z^{c-a}} \int_0^{\infty} e^{-u} u^{c-a-1} \left(1 - \frac{u}{z} \right)^{a-1} du \right.$$

$$\left. + \frac{1}{(-z)^a} \int_0^{\infty} e^{-u} u^{a-1} \left(1 + \frac{u}{z} \right)^{c-a-1} du \right]. \quad (10.41)$$

Because the factor e^z makes the first term much larger than the second term for large values of z, we drop the second term here. From Exercise 8.16 it is clear that for $u > 0$,

$$\left(1 - \frac{u}{z} \right)^{a-1} = \sum_{k=0}^{n-1} \frac{(-a+1)_k}{k! z^k} u^k$$

$$+ \frac{(-a+1)_n}{n! z^n} \left(1 - \frac{u_1}{z} \right)^{a-n-1} u^n, \qquad 0 < u_1 < u.$$

Substituting this result into the first term in Eq. (10.41) yields the approximation

$$_1F_1(a; c; z)$$

$$\approx \frac{\Gamma(c)}{\Gamma(a)\Gamma(c-a)} \frac{e^z}{z^{c-a}} \left[\sum_{k=0}^{n-1} \frac{(-a+1)_k}{k! z^k} \int_0^{\infty} e^{-u} u^{c-a+k-1} du + \Delta_n(z) \right]$$

with

$$\Delta_n(z) = \frac{(-a+1)_n}{n! z^n} \int_0^{\infty} e^{-u} u^{c-a+n-1} \left(1 - \frac{u_1}{z} \right)^{a-n-1} du.$$

For large $z > u_1$ and $\mathrm{Re}(a) > n+1$ it follows that $(1 - u_1/z)^{a-n-1} < 1$. Therefore,

$$|\Delta_n(z)| \leq \left| \frac{(-a+1)_n}{n! z^n} \int_0^{\infty} e^{-u} u^{c-a+n-1} du \right| = \left| \frac{(-a+1)_n}{n! z^n} \Gamma(c+n-a) \right|.$$

Since this last expression goes to zero as $z \to \infty$, we have the asymptotic expansion

$$_1F_1(a; c; z)$$

$$\approx \frac{\Gamma(c)}{\Gamma(a)\Gamma(c-a)} \frac{e^z}{z^{c-a}} \sum_{k=0}^{n-1} \frac{(-a+1)_k}{k! z^k} \int_0^{\infty} e^{-u} u^{c-a+k-1} du, \quad z = |z|.$$

We recognize the integral here as the gamma function. Thus,

$$_1F_1(a; c; z) \approx \frac{\Gamma(c)}{\Gamma(a)\Gamma(c-a)} \frac{e^z}{z^{c-a}} \sum_{k=0}^{n-1} \frac{(-a+1)_k}{k! z^k} \Gamma(c+k-a), \quad z = |z|.$$

$$(10.42)$$

If $z = -|z|$, then for large $|z|$ the factor $e^z = e^{-|z|}$ makes the first term in Eq. (10.41) negligible compared to the second term. In this case

$$_1F_1(a; c; z) \approx \frac{\Gamma(c)}{\Gamma(a)\Gamma(c-a)} \frac{1}{|z|^a} \int_0^\infty e^{-u} u^{a-1} \left(1 - \frac{u}{|z|}\right)^{c-a-1} du$$

$$= \frac{\Gamma(c)}{\Gamma(a)\Gamma(c-a)} \frac{1}{|z|^a} \left[\sum_{k=0}^{n-1} \frac{(a-c+1)_k}{k! |z|^k} \int_0^\infty e^{-u} u^{a+k-1} du + \Delta_n(z)\right].$$

Again the remainder term becomes negligible for large $|z|$, so that

$$_1F_1(a; c; z) \approx \frac{\Gamma(c)}{\Gamma(a)\Gamma(c-a)} \frac{1}{|z|^a} \left[\sum_{k=0}^{n-1} \frac{(a-c+1)_k}{k! |z|^k} \Gamma(a+k)\right], \quad z = -|z|.$$

$$(10.43)$$

Now suppose z is purely imaginary. Let $z = i|z|$. To handle the integral in Eq. (10.40) for this case, consider the contour integral

$$\oint_C e^{tz} t^{a-1} (1-t)^{c-a-1} dt = \int_0^1 e^{tz} t^{a-1} (1-t)^{c-a-1} dt$$

$$+ \int_{C_2+C_3+C_4} e^{tz} t^{a-1} (1-t)^{c-a-1} dt,$$

where C is a rectangular path in the complex plane as shown in Figure 10.5. The integrand is analytic everywhere inside the rectangle for any height h. Thus, the integral over the closed path C vanishes according to Cauchy's theorem and

$$\int_0^1 e^{tz} t^{a-1} (1-t)^{c-a-1} dt = \int_{-C_2-C_3-C_4} e^{tz} t^{a-1} (1-t)^{c-a-1} dt. \quad (10.44)$$

On the parts of the path off the real axis make the following changes of variable:

$$\text{on } C_4: \quad t = is, \qquad 0 < s \le h,$$

$$\text{on } C_3: \quad t = s + ih, \qquad 0 \le s \le 1,$$

$$\text{on } C_2: \quad t = 1 + is, \qquad 0 < s \le h.$$

With these variable changes Eq. (10.44) becomes

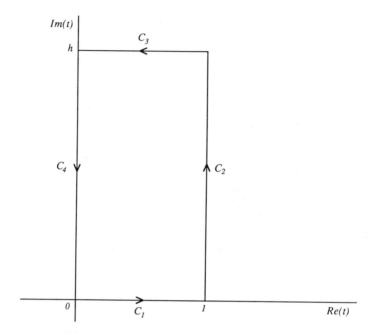

Figure 10.5

$$\int_0^1 e^{tz}t^{a-1}(1-t)^{c-a-1}\,dt = e^{i\frac{1}{2}\pi a}\int_0^h e^{-s|z|}s^{a-1}(1-is)^{c-a-1}\,ds$$

$$+ e^{-h|z|}\int_0^1 e^{is|z|}(s+ih)^{a-1}(1-s-ih)^{c-a-1}\,ds$$

$$+ e^z e^{i\frac{1}{2}\pi(a-c)}\int_0^h e^{-s|z|}(1+is)^{a-1}s^{c-a-1}\,ds.$$

In the limit $h \to \infty$, the second integral vanishes because of the factor $e^{-h|z|}$. If the variable is changed to $u = s|z|$ in the other two integrals and we take the limit $h \to \infty$, then

$$\int_0^1 e^{tz}t^{a-1}(1-t)^{c-a-1}\,dt = \frac{e^{i\frac{1}{2}\pi a}}{|z|^a}\int_0^\infty e^{-u}u^{a-1}\left(1+\frac{u}{z}\right)^{c-a-1}\,du$$

$$+ e^z\frac{e^{i\frac{1}{2}\pi(a-c)}}{|z|^{c-a}}\int_0^\infty e^{-u}\left(1-\frac{u}{z}\right)^{a-1}u^{c-a-1}\,du.$$

In a similar way the result[13] for $z = -i|z|$ can be obtained,

$$\int_0^1 e^{tz} t^{a-1} (1-t)^{c-a-1} dt = \frac{e^{-i\frac{1}{2}\pi a}}{|z|^a} \int_0^\infty e^{-u} u^{a-1} \left(1 + \frac{u}{z}\right)^{c-a-1} du$$

$$+ e^z \frac{e^{i\frac{1}{2}\pi(c-a)}}{|z|^{c-a}} \int_0^\infty e^{-u} \left(1 - \frac{u}{z}\right)^{a-1} u^{c-a-1} du.$$

Make this substitution for the integral in Eq. (10.40) and follow the approximation method above for $z = \pm|z|$ to obtain for purely imaginary $z = \pm i|z|$ the asymptotic formula,

$$_1F_1(a;c;z) \approx \frac{\Gamma(c)}{\Gamma(a)\Gamma(c-a)} \left[\frac{e^{z \mp \frac{1}{2}i\pi(c-a)}}{|z|^{c-a}} \sum_{k=0}^{n-1} \frac{(-a+1)_k}{k! z^k} \Gamma(c+k-a) \right.$$

$$\left. + \frac{e^{\pm \frac{1}{2} i\pi a}}{|z|^a} \sum_{k=0}^{n-1} \frac{(-1)^k (a-c+1)_k}{k! z^k} \Gamma(a+k) \right]. \tag{10.45}$$

By taking only the leading term $(k = 0)$ in each of these cases, the following results[14] are obtained for large values of $|z|$:

$$_1F_1(a;c;z) \approx \frac{\Gamma(c)}{\Gamma(a)} \frac{e^{|z|}}{|z|^{c-a}} \qquad \text{for } z = |z|,$$

$$_1F_1(a;c;z) \approx \frac{\Gamma(c)}{\Gamma(c-a)} \frac{1}{|z|^a} \qquad \text{for } z = -|z|,$$

$$_1F_1(a;c;z) \approx \frac{\Gamma(c)}{\Gamma(a)} \frac{e^{\pm i[|z| - \frac{1}{2}\pi(c-a)]}}{|z|^{c-a}} + \frac{\Gamma(c)}{\Gamma(c-a)} \frac{e^{\pm \frac{1}{2} i\pi a}}{|z|^a} \quad \text{for } z = \pm i|z|.$$

This dependence of the asymptotic behavior on the way in which one approaches infinity is called *Stokes's phenomenon*. It is a general characteristic of asymptotic representations.[15]

These results also lead to the asymptotic formulas for the Bessel functions. For example, the definition of the Bessel function given in Eq. (4.17) yields for real, positive z

$$J_\nu(z) \approx \frac{1}{2} \sqrt{\frac{2}{\pi z}} \sum_{k=0}^{n-1} \frac{(-\nu + \frac{1}{2})_k (\nu + \frac{1}{2})_k}{k! (2iz)^k} \left[e^{i(z - \nu\pi/2 - \pi/4)} \right.$$

$$\left. + (-1)^k e^{-i(z - \nu\pi/2 - \pi/4)} \right]. \tag{10.46}$$

---

[13]In this case the contour in Eq. (10.44) is a rectangle of unit width and height h located in the fourth quadrant of the complex t plane.

[14]*Cf.* Eqs. (3.10).

[15]For a more thorough discussion of this point see P.M. Morse and H. Feshbach, *op. cit.*, p. 609. See also G.N. Watson, *op. cit.*, p. 201.

In arriving at this expression we have used Eq. (10.45) and the gamma function duplication formula.[16]

From the definition of the Neumann function given in Eq. (4.26) we find in a similar fashion that

$$N_\nu(z) \approx \frac{-i}{2} \sqrt{\frac{2}{\pi z}} \sum_{k=0}^{n-1} \frac{\left(-\nu + \frac{1}{2}\right)_k \left(\nu + \frac{1}{2}\right)_k}{k!(2iz)^k}$$

$$\times \left[e^{i(z - \nu\pi/2 - \pi/4)} - (-1)^k e^{-i(z - \nu\pi/2 - \pi/4)} \right]. \qquad (10.47)$$

The results in Eqs. (10.46) and (10.47) lead to the approximation,

$$H_\nu^{(1)}(z) \approx \sqrt{\frac{2}{\pi z}} e^{i(z - \nu\pi/2 - \pi/4)} \sum_{k=0}^{n-1} \frac{\left(-\nu + \frac{1}{2}\right)_k \left(\nu + \frac{1}{2}\right)_k}{k!(2iz)^k}. \qquad (10.48)$$

Lebedev[17] shows that this formula is valid for complex z with $|\arg z| < \pi$. By making use of the result

$$H_\nu^{(2)}(z) = -e^{i\nu\pi} H_\nu^{(1)}(-z)$$

we also have for $|\arg z| < \pi$,

$$H_\nu^{(2)}(z) \approx \sqrt{\frac{2}{\pi z}} e^{-i(z - \nu\pi/2 - \pi/4)} \sum_{k=0}^{n-1} \frac{(-1)^k \left(-\nu + \frac{1}{2}\right)_k \left(\nu + \frac{1}{2}\right)_k}{k!(2iz)^k}. \qquad (10.49)$$

10.9   Continuum States in the Coulomb Field

As an application of the results obtained in the previous section, let us consider the Coulomb scattering of an electrically charged particle. We discussed the bound states for Coulomb attraction in Section 6.4. For scattering $(E > 0)$ we rewrite Eq. (6.30) with slightly different notation as

$$\rho^2 \frac{d^2}{d\rho^2} R(\rho) + 2\rho \frac{d}{d\rho} R(\rho) + \left[\frac{k^2}{\alpha^2}\rho^2 + \lambda\rho - l(l+1) \right] R(\rho) = 0. \qquad (10.50)$$

In this case the wave number $k = \sqrt{2mE/\hbar^2}$ is real. The parameters α and λ are related by[18]

$$\alpha\lambda = -\frac{2mzZe^2}{4\pi\varepsilon_0\hbar^2}, \qquad (10.51)$$

---

[16] See Exercise 10.8.

[17] N.N. Lebedev, *op. cit.*, p. 122. See also G.N. Watson, *op. cit.*, p. 201.

[18] *Cf.* Eq. (6.31). Note that the electron charge $-e$ in Section 6.4 has become ze in the current problem.

where ze is the charge of the scattered particle and Ze the charge of the scatterer. The Coulomb interaction may be attractive or repulsive according to

$$\alpha\lambda > 0, \quad \text{attractive},$$

$$\alpha\lambda < 0, \quad \text{repulsive}.$$

(10.52)

By analogy with the procedure leading to Eq. (6.35) a solution of Eq. (10.50) takes the form

$$R(\rho) = \rho^l e^{q\rho} u(\rho).$$

Making this substitution into Eq. (10.50) and choosing $q = ik/\alpha$ results in the differential equation for $u(\rho)$,

$$\rho u''(\rho) + [2(l+1) + 2ik\rho/\alpha]u'(\rho) + [\lambda + 2ik(l+1)/\alpha]u(\rho) = 0.$$

Because α is as yet undetermined, we can take $\alpha = -2ik$, which gives

$$\rho u''(\rho) + [2(l+1) - \rho]u'(\rho) + [\lambda - (l+1)]u(\rho) = 0$$

with $\rho = -2ikr$. This is the confluent hypergeometric equation. The solution that is regular at the origin is

$$u(\rho) = {}_1F_1\big(l+1+in; 2l+2; -2ikr\big).$$

(10.53)

The parameter $n = i\lambda$ is real. From Eq. (10.52) it is clear that the interaction is attractive for $n < 0$ and repulsive for $n > 0$.

The approximation in Eq. (10.45) leads to the asymptotic form of Eq. (10.53) for large r,

$$u(\rho) \to \frac{2\Gamma(2l+2)}{\Gamma(l+1+in)} \frac{e^{-i(kr-\eta_l)}e^{n\pi/2}}{(2kr)^{l+1}} \sin\big(kr - l\pi/2 - n\log(2kr) + \eta_l\big),$$

where $\eta_l = \arg\Gamma(l+1+in)$. Finally, on putting the pieces together we find that for Eq. (10.50) the asymptotic form of the solution regular at the origin is

$$R(\rho) = \frac{\Gamma(2l+2)}{\Gamma(l+1+in)} \frac{e^{n\pi/2}e^{i(\eta_l - l\pi/2)}}{kr} \sin\big(kr - l\pi/2 - n\log(2kr) + \eta_l\big).$$

EXERCISES

1. A Fraunhofer diffraction pattern is produced when light from a small aperture falls on a screen at a large distance from the aperture. According to Huygens's principle the pattern arises from interference of light arriving at a given point on the screen from different points on the aperture. This is illustrated[19] in Figure 10.6.

---

[19]Figures 10.6 and 10.7 are adapted from E. Hecht and H. Zajac, *Optics*, Addison-Wesley, Reading, MA, 1974 with permission by Addison-Wesley, Inc.

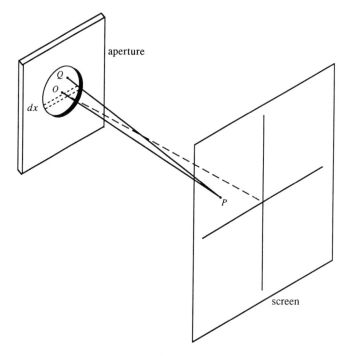

Figure 10.6

The two rays represent light waves from small areas of aperture at the center O and at some arbitrary point Q. For Fraunhofer diffraction the distance r from the aperture to the screen is so large that the rays are considered parallel. Also the radius R of the aperture is so small compared to this distance that we can neglect the effect of the differences in path length on the amplitudes of the light waves at P and consider only their effect on the phases. Assuming monochromatic light of wavelength $\lambda = 2\pi/k$ the optical disturbance $d\psi$ at P due to a small area dA on the aperture is approximately

$$d\psi = C\frac{e^{ik\Delta r}}{r}\,dA$$

where Δr is the difference between the two optical paths QP and OP. The constant C is independent of any position coordinates on the aperture. The total optical disturbance at P is obtained by integrating this expression over the whole aperture.

A typical diffraction pattern for a circular aperture is illustrated in the drawing in Figure 10.7. The characteristic bright central region is called the *Airy disk*. By considering contributions from thin strips of width dx as shown in Figure 10.6 show that the optical disturbance at P can be expressed in terms of a first-order Bessel function, $J_1(\rho)$, as

$$\psi_P = \frac{2\pi R^2 C}{r} \frac{J_1(\rho)}{\rho},$$

with $\rho = kR\sin\theta$. Here θ is the angle between OP and the line from O perpendicular to the screen. Diffraction minima occur at the zeros of $J_1(\rho)$,

$$kR\sin\theta_n = \rho_{1n}, \qquad n = 1, 2, 3, \ldots .$$

Figure 10.7

2. Show by differentiation that the recursion formula

$$\frac{d}{dx}J_\nu(x) = \frac{\nu}{x}J_\nu(x) - J_{\nu+1}(x)$$

follows from Eq. (10.9). In a similar fashion show that

$$\frac{d}{dx}\left[x^{-\nu}J_\nu(x)\right] = -x^{-\nu}J_{\nu+1}(x).$$

3. Follow a procedure similar to that used in obtaining Eq. (10.39) to show that an integral representation of the confluent hypergeometric function is

$$_1F_1(a; c; z) = \frac{\Gamma(c)}{\Gamma(a)\Gamma(c-a)} \int_0^1 e^{tz} t^{a-1}(1-t)^{c-a-1}dt,$$

where $\operatorname{Re}(c) > \operatorname{Re}(a) > 0$.

4. In the integral in the previous exercise change the variable of integration to $u = 1 - t$ to obtain Kummer's transformation[20]

$$_1F_1(a; c; z) = e^z \, _1F_1(c - a; c; -z).$$

5. Use the result in Exercise 10.3 to show that the error function and the incomplete gamma function are related to the confluent hypergeometric function by

$$\text{erf } z = z \, _1F_1 \left(\tfrac{1}{2}; \tfrac{3}{2}; -z^2 \right)$$

and

$$\gamma(a, z) = \frac{z^a}{a} \, _1F_1(a; a + 1; -z),$$

respectively.

6. Two functions that arise in the analysis of Fresnel diffraction by a straight-edge or a rectangular aperture are defined by

$$C(z) = \int_0^z \cos \frac{\pi t^2}{2} \, dt \quad \text{and} \quad S(z) = \int_0^z \sin \frac{\pi t^2}{2} \, dt.$$

These are known as *Fresnel integrals*.[21] Show that each can be expressed in terms of a sum of two confluent hypergeometric functions.

7. From the definition of the Neumann function

$$N_\nu(x) = \frac{J_\nu(x) \cos \nu\pi - J_{-\nu}(x)}{\sin \nu\pi}$$

obtain the integral representation for $N_\nu(x)$ analogous to Eq. (10.10) for $J_\nu(x)$,

$$N_\nu(x) = \frac{1}{\pi} \int_0^\pi \sin(x \sin \theta - \nu\theta) \, d\theta$$
$$- \frac{1}{\pi} \int_0^\infty e^{-x \sinh t} \left(e^{\nu t} + e^{-\nu t} \cos \nu\pi \right) dt.$$

8. Show that for real positive z the approximation

$$J_\nu(z) \approx \frac{1}{2} \sqrt{\frac{2}{\pi z}} \sum_{k=0}^{n-1} \frac{\left(\nu + \tfrac{1}{2} \right)_k \left(\nu + \tfrac{1}{2} \right)_k}{k! (2iz)^k} \left[e^{i(z - \nu\pi/2 - \pi/4)} \right.$$
$$\left. + (-1)^k e^{-i(z - \nu\pi/2 - \pi/4)} \right],$$

---

[20] See also Exercise 9.5.
[21] See, for example, E. Hecht and H. Zajac, *op. cit.*, p. 377.

valid for large values of z, follows from the asymptotic formula for the confluent hypergeometric function. If we consider only the leading terms in the sum, it is clear that

$$J_\nu(z) \xrightarrow[z \to \infty]{} \sqrt{\frac{2}{\pi z}} \cos\left(z - \frac{\nu\pi}{2} - \frac{\pi}{4}\right).$$

Show that this result is consistent with the asymptotic formula for the spherical Bessel function $j_l(x)$ given in Eq. (10.28).

9. From the result in the previous exercise and the definition of the Neumann function show that for z real, large, and positive

$$N_\nu(z) \approx \frac{-i}{2} \sqrt{\frac{2}{\pi z}} \sum_{k=0}^{n-1} \frac{\left(-\nu + \frac{1}{2}\right)_k \left(\nu + \frac{1}{2}\right)_k}{k!(2iz)^k} \left[e^{i(z - \nu\pi/2 - \pi/4)}\right.$$
$$\left. - (-1)^k e^{-i(z - \nu\pi/2 - \pi/4)}\right]$$

from which it follows that

$$N_\nu(z) \xrightarrow[z \to \infty]{} \sqrt{\frac{2}{\pi z}} \sin\left(z - \frac{\nu\pi}{2} - \frac{\pi}{4}\right).$$

10. Obtain the asymptotic formulas for $H_\nu^{(1)}(z)$ and $H_\nu^{(2)}(z)$ given in Eqs. (10.48) and (10.49).

11. For $s = 0$ the differential equation in Exercise 4.6 becomes

$$\frac{d^2}{dx^2} u(x) + \lambda u(x) = 0.$$

Write down the general solution as given in Exercise 4.6 and show that it reduces to the usual solution we obtained in terms of elementary functions when we solved the harmonic oscillator problem in Chapter 1.

12. A particle of mass m and electric charge e is constrained to move in one dimension in a uniform electric field ξ to the right of a perfectly reflecting plane. The potential energy of the particle is given as a function of the position x by

$$V(x) = \begin{cases} \infty, & x < 0, \\ e\xi x, & x > 0. \end{cases}$$

This potential energy function is plotted in Figure 10.8. Set up the time-independent Schrödinger equation for the particle for the region $x > 0$. By redefining the independent variable show that this equation can be written in the form

$$\frac{d^2}{dz^2} u(z) + \lambda z \, u(z) = 0,$$

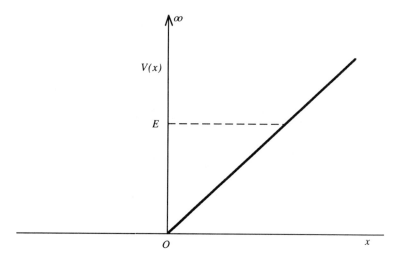

Figure 10.8

which is the equation in Exercise 4.6 with $s = 1$. Write down the general solution to the Schrödinger equation in terms of Hankel functions[22] and then impose the boundary condition

$$u(x) \underset{x \to \infty}{\longrightarrow} 0$$

to obtain the appropriate solution for this particle.

Because the particle cannot penetrate the perfectly reflecting plane at $x = 0$, we also have the boundary condition

$$u(x)\Big|_{x=0} = 0.$$

Use this condition to show that the energy eigenvalue spectrum is discrete with energies given by

$$E_n = \left(\frac{9e^2\xi^2\hbar^2\rho_n^2}{8m}\right)^{1/3}$$

where ρ_n is the position of the nth zero of the Hankel function of the second kind of order $\frac{1}{3}$. That is,

$$H^{(2)}_{\frac{1}{3}}(\rho_n) = 0 \quad \text{with} \quad 0 < \rho_1 < \rho_2 < \rho_3 < \cdots.$$

---

[22]For a solution to this problem in terms of *Airy integrals* see R.G. Winter, *Quantum Physics*, Wadsworth, Belmont, California, 1979, p. 92.

11

Generating Functions and Recursion Formulas

Many authors choose to define the special functions in terms of generating functions. Recursion formulas are also conveniently derived from generating functions. In Chapter 9 we obtained the generating function for the Bessel functions[1]

$$e^{\frac{1}{2}x(u-u^{-1})} = \sum_{k=-\infty}^{\infty} J_k(x)u^k.$$

In this chapter we derive similar expressions for the rest of the special functions we have studied and use them to establish recursion formulas and other relations for these functions.

11.1   Hermite Polynomials

To obtain the generating function for the Hermite polynomials look for a function $g(x,t)$, such that

$$g(x,t) = \sum_{n=0}^{\infty} \frac{H_n(x)}{n!} t^n. \tag{11.1}$$

If $g(x,t)$ is analytic at and near $t = 0$, then Eq. (11.1) is the Taylor series for this function and we have

$$H_n(x) = \left[\frac{\partial^n}{\partial t^n} g(x,t) \right]_{t=0}.$$

Now, $H_n(x)$ is also given by Eq. (9.23), so that

$$(-1)^n e^{x^2} \frac{\partial^n}{\partial x^n} e^{-x^2} = \left[\frac{\partial^n}{\partial t^n} g(x,t) \right]_{t=0}. \tag{11.2}$$

Clearly, if $g(x,t)$ has the form

$$g(x,t) = e^{x^2} f(t-x),$$

---

[1]See Eq. (9.14).

then

$$\left[\frac{\partial^n}{\partial t^n} g(x,t) \right]_{t=0} = \left[e^{x^2} \frac{\partial^n}{\partial t^n} f(t-x) \right]_{t=0}$$

$$= \left[(-1)^n e^{x^2} \frac{\partial^n}{\partial x^n} f(t-x) \right]_{t=0}.$$

On comparison with Eq. (11.2) we see that

$$f(t-x) = e^{-(t-x)^2}.$$

Thus, the generating function for the Hermite polynomials is

$$g(x,t) = e^{x^2} e^{-(t-x)^2} = \sum_{n=0}^{\infty} \frac{H_n(x)}{n!} t^n. \tag{11.3}$$

Since Eq. (11.1) represents the Taylor series of $g(x,t)$ expanded about $t = 0$, then according to Eq. (7.14),

$$H_n(x) = \frac{n!}{2\pi i} \oint_C \frac{e^{x^2} e^{-(u-x)^2}}{u^{n+1}} du. \tag{11.4}$$

The contour C encloses the origin. On rewriting the integrand this reduces to

$$H_n(x) = \frac{n!}{2\pi i} \oint_C \frac{e^{2xu-u^2}}{u^{n+1}} du. \tag{11.5}$$

In this form it is easy to see that

$$H_n'(x) = \frac{n!}{2\pi i} \oint_C \frac{2u e^{2xu-u^2}}{u^{n+1}} du$$

$$= 2n \frac{(n-1)!}{2\pi i} \oint_C \frac{e^{2xu-u^2}}{u^{(n-1)+1}} du$$

$$= 2n H_{n-1}(x).$$

So, we arrive at the formula,

$$H_n'(x) = 2n H_{n-1}(x). \tag{11.6}$$

Now let us take the derivative of each side of this equation and invoke Eq. (11.6) again. This gives

$$H_n''(x) = 4n(n-1) H_{n-2}(x). \tag{11.7}$$

Substituting from Eqs. (11.6) and (11.7) into Hermite's equation (Eq. 3.14) and redefining indices we obtain the recursion formula,

$$H_{n+1}(x) = 2x H_n(x) - 2n H_{n-1}(x). \tag{11.8}$$

The first few of the Hermite polynomials are given in Table 11.1.

TABLE 11.1. Hermite polynomials

n	$H_n(x)$
0	1
1	$2x$
2	$4x^2 - 2$
3	$8x^3 - 12x$
4	$16x^4 - 48x^2 + 12$
5	$32x^5 - 160x^3 + 120x$
6	$64x^6 - 480x^4 + 720x^2 - 120$

11.2 Laguerre Polynomials

11.2.1 THE GENERATING FUNCTION

The generating function for the Laguerre polynomials is easily obtained from Eq. (6.44). First, look for a function $w(x, t)$, such that

$$w(x, t) = \sum_{k=0}^{\infty} \frac{L_k(x)}{k!} t^k. \tag{11.9}$$

This is simply the Taylor expansion of an analytic function of t near $t = 0$. Therefore, $L_k(x)$ can be expressed as a contour integral,

$$L_k(x) = \frac{k!}{2\pi i} \oint_C \frac{w(x, t)}{t^{k+1}} \, dt. \tag{11.10}$$

From Eq. (6.44) we also have the formula

$$L_k(x) = e^x \frac{d^k}{dx^k} (x^k e^{-x}).$$

According to Eq. (7.14), which gives the nth derivative of an analytic function, we can write this last expression as

$$L_k(x) = \frac{k!}{2\pi i} \oint_C \frac{u^k e^{-(u-x)}}{(u-x)^{k+1}} \, du$$

$$= \frac{k!}{2\pi i} \oint_C \left(\frac{u}{u-x} \right)^k \frac{e^{-(u-x)}}{u-x} \, du. \tag{11.11}$$

A comparison of the integrand here with that in Eq. (11.10) suggests a variable change in Eq. (11.11) to $t = (u - x)/u$, which gives

$$L_k(x) = \frac{k!}{2\pi i} \oint_C \frac{e^{-xt/(1-t)}}{(1-t)t^{k+1}} \, dt. \tag{11.12}$$

On comparison with Eq. (11.10) we see that

$$w(x,t) = \frac{e^{-xt/(1-t)}}{1-t} \tag{11.13}$$

is the generating function for the Laguerre polynomials.

11.2.2 RECURSION RELATIONS FOR LAGUERRE POLYNOMIALS

The generating function is useful in obtaining recursion formulas for the Laguerre polynomials. By differentiating Eq. (11.13) with respect to t we find that

$$(1-t)^2 \frac{\partial w(x,t)}{\partial t} + (x+t-1)w(x,t) = 0.$$

Substitute into this equation the series for $w(x,t)$ from Eq. (11.9) and collect terms to obtain

$$\big[(x-1)L_0(x) + L_1(x)\big]t^0$$

$$+ \sum_{k=1}^{\infty} \frac{1}{k!}\big[L_{k+1}(x) + (x-1-2k)L_k(x) + k^2 L_{k-1}(x)\big]t^k = 0.$$

Since this relation holds for arbitrary t, we must have

$$L_1(x) = (1-x)L_0(x)$$

and

$$L_{k+1}(x) + (x-1-2k)L_k(x) + k^2 L_{k-1}(x) = 0, \qquad k > 0. \tag{11.14}$$

Starting with $L_0(x) = 1$,[2] we can generate all higher-order Laguerre polynomials from these relations. The first few are tabulated in Table 11.2.

Similarly, differentiation of Eq. (11.13) with respect to x yields

$$(1-t)\frac{\partial w(x,t)}{\partial x} + tw(x,t) = 0.$$

Now insert into this equation the expansion for $w(x,t)$ from Eq. (11.9) and collect terms to get

$$\sum_{k=1}^{\infty} \frac{1}{k!}\left[\frac{d}{dx}L_k(x) - k\frac{d}{dx}L_{k-1}(x) + kL_{k-1}(x)\right]t^k = 0,$$

which holds for arbitrary t. Thus,

$$\frac{d}{dx}L_k(x) - k\frac{d}{dx}L_{k-1}(x) + kL_{k-1}(x) = 0, \qquad k > 0. \tag{11.15}$$

---

[2] For example, from Eq. (6.44), it is clear that $L_0(x) = 1$.

TABLE 11.2. Laguerre polynomials

n	$L_n(x)$
0	1
1	$1 - x$
2	$2 - 4x + x^2$
3	$6 - 18x + 9x^2 - x^3$
4	$24 - 96x + 72x^2 - 16x^3 + x^4$
5	$120 - 600x + 600x^2 - 200x^3 + 25x^4 - x^5$
6	$720 - 4320x + 5400x^2 - 2400x^3 + 450x^4 - 36x^5 + x^6$

This result may also be obtained directly from differentiation of the Rodrigues formula (Eq. 6.44).[3]

The confluent hypergeometric function in Eq. (6.44) also yields a recursion relation for the derivative of $L_k(x)$. On differentiating the confluent hypergeometric series term by term, we get

$$\frac{d}{dx}L_k(x) = \frac{k!}{x}\sum_{m=0}^{\infty}\frac{(-k)_m}{m!m!}(m - k + k)x^m.$$

Now make use of the identity

$$(-k)_m(-k + m) = -k(-k + 1)_m$$

to write this equation as

$$\frac{d}{dx}L_k(x) = \frac{k!}{x}\left[-k\sum_{m=0}^{\infty}\frac{(-k + 1)_m}{m!m!}x^m + k\sum_{m=0}^{\infty}\frac{(-k)_m}{m!m!}x^m\right].$$

From this expression it follows that

$$x\frac{d}{dx}L_k(x) = -k^2 L_{k-1}(x) + kL_k(x). \tag{11.16}$$

11.3 Associated Laguerre Polynomials

11.3.1 THE GENERATING FUNCTION

We can derive the generating function for the associated Laguerre polynomials from the one for the Laguerre polynomials. Let us differentiate both sides of Eq. (11.9) p times with respect to x,

$$\frac{\partial^p}{\partial x^p}w(x,t) = \sum_{k=0}^{\infty}\frac{1}{k!}\left[\frac{d^p}{dx^p}L_k(x)\right]t^k = \sum_{k=0}^{\infty}\frac{L_k^p(x)}{k!}t^k. \tag{11.17}$$

---

[3]See Exercise 11.3.

Clearly, $(\partial^p / \partial x^p) w(x, t)$ is the generating function for the $L_k^p(x)$.

If this differentiation is carried out explicitly in Eq. (11.13), the generating function for the associated Laguerre polynomials is found to be

$$(-1)^p \frac{t^p}{(1-t)^{p+1}} e^{-xt/(1-t)} = \sum_{q=0}^{\infty} \frac{L_q^p(x)}{q!} t^q. \qquad (11.18)$$

Starting with this result we can derive the Rodrigues formula for the associated Laguerre polynomials (Eq. 9.25). Since the left-hand side of Eq. (11.18) is a function of t analytic near $t = 0$, $L_k^p(x)$ can be expressed as a contour integral,

$$L_q^p(x) = (-1)^p \frac{q!}{2\pi i} \oint_C \frac{e^{-xt/(1-t)}}{(1-t)^{p+1} t^{q-p+1}} dt. \qquad (11.19)$$

With the change of variable $u = x/(1-t)$ this expression becomes

$$L_q^p(x) = (-1)^p e^x x^{-p} \frac{q!}{2\pi i} \oint_C \frac{e^{-u} u^q}{(u-x)^{q-p+1}} du. \qquad (11.20)$$

According to Eq. (7.14) the integral here is proportional to the $(q-p)$th derivative of $e^{-x} x^q$. Thus,

$$L_q^p(x) = (-1)^p \frac{q!}{(q-p)!} e^x x^{-p} \frac{d^{q-p}}{dx^{q-p}} (e^{-x} x^q). \qquad (11.21)$$

The difference $q-p$ is a nonnegative integer, but q and p themselves are not necessarily integers. In this case the more general form for the associated Laguerre function[4] is

$$L_q^p(x) = (-1)^p \frac{\Gamma(q+1)}{(q-p)!} e^x x^{-p} \frac{d^{q-p}}{dx^{q-p}} (e^{-x} x^q), \qquad (11.22)$$

which is Eq. (9.25).

11.3.2 Recursion Formulas for Associated Laguerre Polynomials

Now use the generating function to obtain recursion relations for the associated Laguerre polynomials. First, differentiate the generating function in Eq. (11.18) to get

$$t(1-t)^2 \frac{\partial}{\partial t} \left[(-1)^p \frac{t^p}{(1-t)^{p+1}} e^{-xt/(1-t)} \right]$$

$$+ \left(t^2 - (1-p-x)t - p \right) \left[(-1)^p \frac{t^p}{(1-t)^{p+1}} e^{-xt/(1-t)} \right] = 0.$$

---

[4]See Exercise 11.5.

On substituting the series expansion for the generating function from Eq. (11.18), this equation becomes

$$-pL_0^p(x)t^0 + \left[(x+p-1)L_0^p(x) + (1-p)L_1^p(x)\right]t^1$$

$$+ \sum_{k=2}^{\infty} \frac{1}{k!}\Big[(k-p)L_k^p(x)$$

$$+ k(x+p+1-2k)L_{k-1}^p(x) + k(k-1)^2 L_{k-2}^p(x)\Big]t^k = 0.$$

Again we use the linear independence of terms in different powers of t and arrive at the recursion formula,

$$(k+1-p)L_{k+1}^p(x) + (k+1)(x+p-2k-1)L_k^p(x)$$

$$+ k^2(k+1)L_{k-1}^p(x) = 0, \qquad k > 0. \qquad (11.23)$$

Another recursion relation follows from the identity

$$(-1)^p \frac{t^p}{(1-t)^{p+1}} e^{-xt/(1-t)} = \frac{-t}{1-t}\left[(-1)^{p-1}\frac{t^{p-1}}{(1-t)^p} e^{-xt/(1-t)}\right].$$

First, substitute the series in Eq. (11.18) for the generating function. Then, by collecting and rearranging terms we find that

$$L_0^p(x)t^0 + \sum_{k=1}^{\infty} \frac{1}{k!}\left[L_k^p(x) - kL_{k-1}^p(x) + kL_{k-1}^{p-1}(x)\right]t^k = 0.$$

Since this must hold for arbitrary t,

$$L_k^p(x) - kL_{k-1}^p(x) + kL_{k-1}^{p-1}(x) = 0, \qquad k \geq p > 0. \qquad (11.24)$$

A combination of Eqs. (11.23) and (11.24) leads to the formula,

$$(k-p+1)L_{k+1}^p(x) + (k+1)(x+p-k-1)L_k^p(x) + k^2(k+1)L_{k-1}^{p-1}(x) = 0. \qquad (11.25)$$

We can also differentiate the generating function with respect to x,

$$\frac{\partial}{\partial x}\left[(-1)^p \frac{t^p}{(1-t)^{p+1}} e^{-xt/(1-t)}\right] = (-1)^{p+1}\frac{t^{p+1}}{(1-t)^{p+2}} e^{-xt/(1-t)}.$$

On substituting the series representation for the generating function from Eq. (11.18) into this equation and equating coefficients of equal powers of t, we get

$$\frac{d}{dx}L_k^p(x) = L_k^{p+1}(x). \qquad (11.26)$$

This relation also follows directly from Eq. (6.45)[5] and it can be used to derive easily the associated Laguerre polynomials from the Laguerre polynomials. A few of these are given in Table 11.3.

---

[5]See Exercise 11.4.

TABLE 11.3. Associated Laguerre polynomials

k	p	$L_k^p(x)$
0	0	1
1	0	$1 - x$
	1	-1
2	0	$2 - 4x + x^2$
	1	$-4 + 2x$
	2	2
3	0	$6 - 18x + 9x^2 - x^3$
	1	$-18 + 18x - 3x^2$
	2	$18 - 6x$
	3	-6
4	0	$24 - 96x + 72x^2 - 16x^3 + x^4$
	1	$-96 + 144x - 48x^2 + 4x^3$
	2	$144 - 96x + 12x^2$
	3	$-96 + 24x$
	4	24
5	0	$120 - 600x + 600x^2 - 200x^3 + 25x^4 - x^5$
	1	$-600 + 1200x - 600x^2 + 100x^3 - 5x^4$
	2	$1200 - 1200x + 300x^2 - 20x^3$
	3	$-1200 + 600x - 60x^2$
	4	$600 - 120x$
	5	-120

Another useful derivative formula is obtained by differentiation of Eq. (6.47),

$$\frac{d}{dx} L_q^p(x) = \frac{(-1)^p [\Gamma(q+1)]^2}{x\Gamma(p+1)\Gamma(q-p+1)} \sum_{m=0}^{\infty} \frac{m(-q+p)_m}{m!(p+1)_m} x^m.$$

Using the identity

$$(-q+p)_m m = (-q+p)_m(-q+p+m+q-p)$$
$$= (-q+p)[(-q+p+1)_m - (-q+p)_m],$$

rewrite this equation as

$$x\frac{d}{dx} L_q^p(x) = (q-p)L_q^p(x) - q^2 L_{q-1}^p(x). \tag{11.27}$$

11.4   Legendre Polynomials

11.4.1   The Generating Function

Schläfli's integral (Eq. 10.30) can be used to obtain the generating function for the Legendre polynomials. Look for a function $g(x, u)$, such that

$$g(x, u) = \sum_{n=0}^{\infty} P_n(x) u^n.$$

Clearly, $g(x, u)$ is an analytic function of u near $u = 0$ with the coefficients in the Taylor expansion given by

$$P_n(x) = \frac{1}{2\pi i} \oint_C \frac{g(x, u)}{u^{n+1}} \, du. \tag{11.28}$$

We require that $P_n(x)$ also be given by Eq. (10.30), which we rewrite as

$$P_n(x) = \frac{1}{\pi i} \oint_{C'} \left(\frac{t^2 - 1}{2(t - x)} \right)^{n+1} \frac{dt}{t^2 - 1}. \tag{11.29}$$

A comparison of the integrands in Eqs. (11.28) and (11.29) suggests a change of variable in Eq. (11.29) to

$$u = \frac{2(t - x)}{t^2 - 1}$$

or

$$t = u^{-1} \left[1 \pm \sqrt{1 - 2xu + u^2} \right].$$

If we choose the lower sign, Eq. (11.29) becomes

$$P_n(x) = \frac{1}{2\pi i} \oint_C \frac{(1 - 2xu + u^2)^{-1/2}}{u^{n+1}} \, du. \tag{11.30}$$

Finally, compare Eq. (11.28) with Eq. (11.30); the generating function for $P_n(x)$ is then seen to be

$$g(x, u) = \left[1 - 2xu + u^2 \right]^{-1/2} = \sum_{n=0}^{\infty} P_n(x) u^n. \tag{11.31}$$

11.4.2   Normalization Integral

The normalization integral for the Legendre polynomials follows from Eq. (11.31). First, square both sides of this equation to get

$$\frac{1}{1 - 2xu + u^2} = \sum_{m=0}^{\infty} \sum_{n=0}^{\infty} P_m(x) P_n(x) u^{m+n}.$$

Now integrate from -1 to $+1$ and use the relation for Legendre polynomials obtained in Eq. (5.24). This gives,

$$\int_{-1}^{1} \frac{dx}{1 - 2xu + u^2} = \sum_{n=0}^{\infty} \left\{ \int_{-1}^{1} [P_n(x)]^2 dx \right\} u^{2n}. \tag{11.32}$$

For the left-hand side,

$$\frac{\log[1 - 2xu + u^2]}{-2u} \bigg|_{-1}^{1} = \frac{1}{u} \left[\log(1 + u) - \log(1 - u) \right].$$

From Eq. (2.7) we find that for $|u| < 1$

$$\log(1 + u) - \log(1 - u) = \sum_{k=1}^{\infty} \frac{(-1)^{k-1} + 1}{k} u^k. \tag{11.33}$$

Since

$$(-1)^{k-1} + 1 = \begin{cases} 2, & k = \text{odd}, \\ 0, & k = \text{even}, \end{cases}$$

it is clear that only odd powers of u appear in the sum in Eq. (11.33). Redefine the summation index to reflect this fact. Then,

$$\log(1 + u) - \log(1 - u) = 2 \sum_{n=0}^{\infty} \frac{u^{2n+1}}{2n + 1}.$$

With this result the left-hand side of Eq. (11.32) can be evaluated. After some rearrangement Eq. (11.32) becomes

$$\sum_{n=0}^{\infty} \left\{ \frac{2}{2n + 1} - \int_{-1}^{1} [P_n(x)]^2 dx \right\} u^{2n} = 0.$$

Since this holds for arbitrary u, we must have

$$\int_{-1}^{1} [P_n(x)]^2 dx = \frac{2}{2n + 1}.$$

By combining this result with Eq. (5.24) we obtain the property

$$\int_{-1}^{1} P_m(x) P_n(x) \, dx = \frac{2}{2n + 1} \delta_{mn}, \tag{11.34}$$

for the Legendre polynomials. The *Kronecker delta* symbol δ_{mn} is defined by

$$\delta_{mn} = \begin{cases} 1, & m = n, \\ 0, & m \neq n. \end{cases} \tag{11.35}$$

In the next chapter we derive a relation similar to Eq. (11.34) for the associated Legendre functions.

11.4.3 RECURSION FORMULAS FOR LEGENDRE POLYNOMIALS

Differentiation of the generating function with respect to x gives

$$\frac{\partial}{\partial x}\, g(x,t) = \frac{t}{1 - 2xt + t^2}\, g(x,t). \tag{11.36}$$

Now substitute the series expansion for $g(x,t)$ from Eq. (11.31) into Eq. (11.36). This leads to the equation,

$$P_0'(x) + \big[P_1'(x) - 2xP_0'(x) - P_0(x)\big]t$$
$$+ \sum_{n=2}^{\infty}\big[P_n'(x) - 2xP_{n-1}'(x) + P_{n-2}'(x) - P_{n-1}(x)\big]t^n = 0.$$

Set the coefficients of separate powers of t equal to zero to get the formulas,

$$P_0'(x) = 0,$$

$$P_1'(x) - P_0(x) = 0, \tag{11.37}$$

$$P_{n+1}'(x) - 2xP_n'(x) + P_{n-1}'(x) - P_n(x) = 0.$$

Other recursion relations also follow from the generating function. Differentiation of $g(x,t)$ with respect to t yields

$$\frac{\partial}{\partial t}\, g(x,t) = \frac{x - t}{1 - 2xt + t^2}\, g(x,t). \tag{11.38}$$

Combining Eqs. (11.36) and (11.38) leads to

$$(x - t)\frac{\partial}{\partial x}\, g(x,t) - t\frac{\partial}{\partial t}\, g(x,t) = 0. \tag{11.39}$$

On substituting the series expansion for $g(x,t)$ into Eq. (11.39) we find after some rearrangement that

$$xP_0'(x) + \sum_{n=1}^{\infty}\big[xP_n'(x) - P_{n-1}'(x) - nP_n(x)\big]t^n = 0.$$

Again the coefficients of different powers of t must vanish separately. Hence, the relations,

$$P_0'(x) = 0,$$

$$\tag{11.40}$$

$$xP_n'(x) - P_{n-1}'(x) - nP_n(x) = 0, \qquad n > 0,$$

follow.

Substitution from Eq. (11.40) for $P'_{n-1}(x)$ into Eq. (11.37) gives the formula

$$P'_{n+1}(x) - xP'_n(x) - (n+1)P_n(x) = 0. \tag{11.41}$$

This equation can also be written as

$$P'_n(x) - xP'_{n-1}(x) - nP_{n-1}(x) = 0, \qquad n > 0. \tag{11.42}$$

Finally, substituting $P'_{n-1}(x)$ from Eq. (11.40) into Eq. (11.42) yields the formula

$$(1 - x^2)P'_n(x) - nP_{n-1}(x) + nxP_n(x) = 0. \tag{11.43}$$

We can also obtain by combining Eqs. (11.40) and (11.42) a very convenient formula for deriving the first n Legendre polynomials,

$$(x^2 - 1)P'_{n-1}(x) + nxP_{n-1}(x) - nP_n(x) = 0, \qquad n > 0. \tag{11.44}$$

This relation was used to construct Table 11.4.

TABLE 11.4. Legendre polynomials

n	$P_n(x)$
0	1
1	x
2	$\frac{1}{2}(3x^2 - 1)$
3	$\frac{1}{2}(5x^3 - 3x)$
4	$\frac{1}{8}(35x^4 - 30x^2 + 3)$
5	$\frac{1}{8}(63x^5 - 70x^3 + 15x)$
6	$\frac{1}{16}(231x^6 - 315x^4 + 105x^2 - 5)$

11.5 Associated Legendre Functions

To obtain the generating function for the associated Legendre functions we start by differentiating the generating function for $P_n(x)$ $|m|$ times with respect to x. We find that

$$\frac{\partial^{|m|}}{\partial x^{|m|}}\left[1 - 2xt + t^2\right]^{-1/2} = \frac{\left(\frac{1}{2}\right)_{|m|}(2t)^{|m|}}{\left[1 - 2xt + t^2\right]^{|m|+\frac{1}{2}}}.$$

With this result if Eq. (11.31) is differentiated $|m|$ times with respect to x and multiplied by $(1 - x^2)^{\frac{1}{2}|m|}$, then

$$\frac{(1 - x^2)^{\frac{1}{2}|m|} \left(\frac{1}{2}\right)_{|m|} (2t)^{|m|}}{\left[1 - 2xt + t^2\right]^{|m|+\frac{1}{2}}} = \sum_{n=0}^{\infty} (1 - x^2)^{\frac{1}{2}|m|} \frac{d^{|m|}}{dx^{|m|}} P_n(x) \, t^n.$$

On comparing this equation with Eq. (5.22) we recognize the associated Legendre function $P_n^m(x)$ on the right-hand side here. Thus, the generating function for $P_n^m(x)$ is

$$\frac{(2|m|)!}{2^{|m|}|m|!} \frac{(1 - x^2)^{\frac{1}{2}|m|} t^{|m|}}{\left[1 - 2xt + t^2\right]^{|m|+\frac{1}{2}}} = \sum_{n=0}^{\infty} P_n^m(x) t^n. \tag{11.45}$$

In arriving at this expression, we have used the identity

$$(2k)! = 2^{2k} \left(\tfrac{1}{2}\right)_k k! \, .$$

11.6 A Note on Normalization

Any polynomial function of z is analytic everywhere in the complex plane. It is clear that for any set of polynomials $T_n(z)$ defined by a generating function $w(z, t)$ according to either

$$w(z, t) = \sum_{n=0}^{\infty} T_n(z) t^n$$

or

$$w(z, t) = \sum_{n=0}^{\infty} \frac{T_n(z)}{n!} t^n,$$

the normalization of the polynomials must be such that

$$T_0(z) = w(z, 0).$$

We have used this property to get the starting functions to produce Tables 11.1 to 11.4 from the recursion formulas.

EXERCISES

1. Derive the recursion formula (Eq. 4.21) for the Bessel functions by differentiating with respect to t both sides of the equation defining the generating function,

$$e^{\frac{1}{2}z(t - t^{-1})} = \sum_{n=-\infty}^{\infty} J_n(z) \, t^n.$$

2. Use the generating function in Exercise 11.1 to show that

$$J_n(x+y) = \sum_{m=-\infty}^{\infty} J_m(x) J_{n-m}(y).$$

Refer to Exercise (2.26).

3. Derive the recursion relation for derivatives of the Laguerre polynomials in Eq. (11.15) from Rodrigues's formula (Eq. 6.44).

4. Show that the formula

$$\frac{d}{dx} L_k^p(x) = L_k^{p+1}(x)$$

follows immediately from Eq. (6.45).

5. Use Leibniz's theorem to prove that

$$\frac{d^{q-p}}{dx^{q-p}} (e^{-x} x^q) = \frac{\Gamma(q+1)}{\Gamma(p+1)} x^p e^{-x} \, {}_1F_1(-q+p; p+1; x),$$

where q and p need not be integers, but

$$q - p = 0, 1, 2, 3, \ldots .$$

By comparison with Eq. (6.47) show that Eq. (11.22), the Rodrigues formula for associated Laguerre functions, follows.

6. Fill in the details of the derivation from Schläfli's integral of the generating function for Legendre polynomials in Eq. (11.31).

7. The generating function for the Legendre polynomials is

$$\left[1 - 2zt + t^2\right]^{-1/2} = \sum_{n=0}^{\infty} P_n(z) t^n.$$

Differentiate both sides of this equation with respect to t to obtain the recursion formula

$$(n+1) P_{n+1}(z) - (2n+1) z P_n(z) + n P_{n-1}(z) = 0, \qquad n > 0.$$

8. Use Laplace's integral representation,

$$P_n(z) = \frac{1}{\pi} \int_0^{\pi} \left(z + \sqrt{z^2 - 1} \cos \phi\right)^n d\phi,$$

to derive the generating function for Legendre polynomials. The geometric series and the result of Exercise 8.6 are useful here.

9. At a distance s from a point charge q the electric potential U is given by

$$U = \frac{1}{4\pi\varepsilon_0}\frac{q}{s}.$$

Suppose the charge is located on the z axis at a distance b from the origin as shown in Figure 11.1. Use the generating function for Legendre polynomials to show that at points such that $r > b$ the potential is

$$U(r,\theta) = \frac{q}{4\pi\varepsilon_0}\sum_{n=0}^{\infty} P_n(\cos\theta)\frac{b^n}{r^{n+1}}.$$

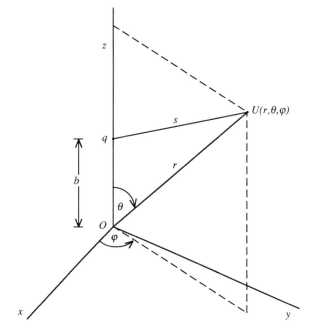

Figure 11.1

10. A polynomial of degree n is defined by

$$S_n(x) = \sum_{m=0}^{n}\frac{x^m}{m!}.$$

Derive a generating function $w(x,t)$, such that

$$w(x,t) = \sum_{n=0}^{\infty} S_n(x)\,t^n.$$

From this generating function obtain the recursion formulas

$$\frac{d}{dx}S_n(x) = S_{n-1}(x) \quad \text{for} \ \ n > 0,$$

and

$$(n+1)S_{n+1}(x) - (n+1+x)S_n(x) + xS_{n-1}(x) = 0 \quad \text{for} \ \ n > 0.$$

11. Express the polynomial $S_n(x)$ in Exercise 11.10 as a contour integral in the complex plane.

12

Orthogonal Functions

12.1 Complete Orthonormal Sets of Functions

In quantum mechanics, as well as other branches of physics, it is convenient to deal with *complete* sets of *orthonormal* functions. By orthonormal we mean that the functions have the property[1]

$$\int u_n^*(z)u_m(z)\,dz = \delta_{mn}. \tag{12.1}$$

A set of functions $\{u_n(z)\}$ is said to be complete if any well-behaved function $f(z)$ can be expressed as a linear superposition of these functions,[2]

$$f(z) = \sum_k c_k u_k(z), \tag{12.2}$$

where the coefficients c_k are independent of z. The label k could be continuous instead of discrete in which case we write

$$f(z) = \int c(k)u(k,z)\,dk. \tag{12.3}$$

If the functions $u_k(z)$ are orthonormal in the sense of Eq. (12.1), then the evaluation of the c_k becomes particularly simple. We multiply Eq. (12.2) by $u_m^*(z)$ and integrate over the complete range of z. This yields the result

$$\int u_m^*(z)f(z)\,dz = \sum_k c_k \int u_m^*(z)u_k(z)\,dz, \tag{12.4}$$

where the order of summation and integration is reversed. Using the orthonormality relation of Eq. (12.1) in Eq. (12.4) yields

$$c_m = \int u_m^*(z)f(z)\,dz. \tag{12.5}$$

Substitute this result into Eq. (12.2) for

$$f(z) = \int f(z')\left[\sum_k u_k^*(z')u_k(z)\right]dz'.$$

---

[1]For real $u_n(z)$ complex conjugation is of no consequence.
[2]In general, this is an infinite series, which we assume converges.

We assume that $f(z')$ is a local function, by which we mean that its value at $z' = z$ does not depend on its value at some other point $z' \neq z$. Thus,

$$\sum_k u_k^*(z')u_k(z) = \delta(z', z), \qquad (12.6a)$$

where $\delta(z', z)$ is a generalization to continuous variables z and z' of the Kronecker delta defined in Eq. (11.35) for the discrete variables m and n. If k is continuous, Eq. (12.6a) takes the form

$$\int u^*(z', k)u(z, k)\, dk = \delta(z', z). \qquad (12.6b)$$

The relation in Eqs. (12.6) is the *completeness* or *closure* condition.

12.2   The Delta Function

To obtain the appropriate generalization[3] for $\delta(z', z)$ we use an alternate definition of the Kronecker delta provided by

$$A_n = \sum_m \delta_{mn} A_m.$$

For a function of a continuous variable $f(z)$ the generalization is

$$f(z) = \int_{-\infty}^{\infty} f(z')\delta(z', z)\, dz'. \qquad (12.7)$$

We assume that $f(z)$ does not depend on values of $z' \neq z$. This means that

$$\delta(z', z) = 0, \qquad z' \neq z. \qquad (12.8)$$

According to Eq. (12.7)

$$f(z + a) = \int_{-\infty}^{\infty} f(z' + a)\delta(z' + a, z + a)\, dz'.$$

Also consistent with the properties of $\delta(z', z)$ implied in Eq. (12.7) is the statement

$$f(z + a) = \int_{-\infty}^{\infty} f(z' + a)\delta(z', z)\, dz'.$$

On comparing these last two expressions it is clear that $\delta(z' + a, z + a)$ and $\delta(z', z)$ are equivalent. Since a is arbitrary, take $a = -z$, in which case $\delta(z', z)$ depends only on the *difference* $z' - z$. To make this explicit, write

$$\delta(z', z) = \delta(z' - z). \qquad (12.9)$$

---

[3]This development through Eq. (12.13) follows E. Merzbacher, *op. cit.*, p. 81.

With this notation Eq. (12.7) becomes

$$f(z) = \int_{-\infty}^{\infty} f(z')\delta(z' - z)\, dz'. \tag{12.10}$$

On rewriting Eq. (12.8) we have

$$\delta(z' - z) = 0, \qquad z' \neq z. \tag{12.11}$$

From Eqs. (12.10) and (12.11) it follows that

$$\int_{-\infty}^{\infty} \delta(z' - z)\, dz' = 1. \tag{12.12}$$

If $f(z)$ is an odd function of z then $f(0) = 0$. In this case

$$f(0) = \int_{-\infty}^{\infty} f(t)\delta(t)\, dt = 0,$$

which requires that $\delta(t)$ be an even function, that is,

$$\delta(-t) = \delta(t). \tag{12.13}$$

Taken together Eqs. (12.11) and (12.12) define an entity that becomes infinite in a manner such that the area under the curve $y = \delta(t)$ is equal to one. It was used by Dirac in his classic text on quantum mechanics[4] and is known as the *Dirac delta function*.

Strictly speaking $\delta(t)$ is not a function at all in the usual sense. However, since it is closely approximated by a very narrow spike of unit area and is symmetric about $t = 0$, we may represent $\delta(t)$ by such a function in the limit that the width goes to zero while the area remains equal to one. Some examples of functions, denoted by $y(t, p)$, meeting these criteria are plotted in Figure 12.1. The parameter p is a measure of the sharpness of the spike, hence the definition

$$\delta(t) = \lim_{p \to \infty} y(t, p). \tag{12.14}$$

The oscillating function

$$y(t, p) = \frac{\sin pt}{\pi t} \tag{12.15}$$

is plotted in Figure 12.2 for several values of p. We see that as p increases, this function becomes more and more sharply peaked around $t = 0$. The zeros of $y(t, p)$ as a function of t occur at intervals of π/p with those nearest

---

[4]P.A.M. Dirac, *Principles of Quantum Mechanics*, Oxford University Press, Oxford, 1928, p. 58.

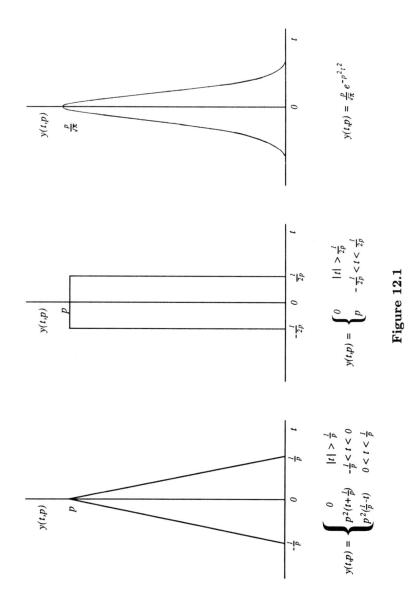

Figure 12.1

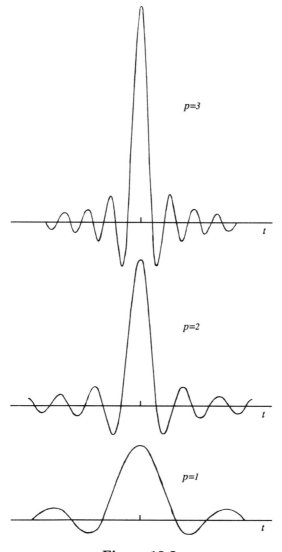

$p=3$

$p=2$

$p=1$

Figure 12.2

$t = 0$ occurring at $t = \pm\pi/p$. At $t = 0$ the value of $y(t, p)$ is p/π. So, in the limit $p \to \infty$, $y(t, p)$ becomes an infinitely sharp spike.

Suppose $y(t, p)$ is plotted as a function of t for any value of p. The area under the curve is given by the integral

$$\int_{-\infty}^{\infty} y(t, p)\, dt = \int_{-\infty}^{\infty} \frac{\sin pt}{\pi t}\, dt.$$

Now apply Jordan's lemma and take the Cauchy principal value,[5] which yields

$$\int_{-\infty}^{\infty} \frac{\sin pt}{\pi t} \, dt = 1.$$

Therefore, another suitable representation of the delta function is

$$\delta(t) = \lim_{p \to \infty} \frac{\sin pt}{\pi t}. \tag{12.16}$$

The function defined by Eq. (12.15) is equal to the definite integral

$$\frac{1}{2\pi} \int_{-p}^{p} e^{itx} \, dx = \frac{\sin pt}{\pi t}.$$

This expression, coupled with Eq. (12.16), provides an integral representation of the Dirac delta function,

$$\delta(t) = \frac{1}{2\pi} \int_{-\infty}^{\infty} e^{itx} \, dx. \tag{12.17}$$

From this result it is seen that the functions $u(z, k) = \frac{1}{\sqrt{2\pi}} e^{ikz}$ satisfy the completeness relation, because

$$\frac{1}{2\pi} \int_{-\infty}^{\infty} e^{ik(z'-z)} \, dk = \delta(z' - z),$$

as required by Eq. (12.6b). Thus, the elementary functions $e^{\pm ikx}$, as well as $\sin kx$ and $\cos kx$, form complete sets of functions.

12.3   Fourier Integrals

The integral expansion of an arbitrary function $f(x)$ in terms of the functions $e^{\pm ikx}$ is known as a *Fourier integral*,

$$f(x) = \frac{1}{\sqrt{2\pi}} \int_{-\infty}^{\infty} g(k) e^{ikx} \, dk. \tag{12.18}$$

The function $g(k)$, called the *Fourier amplitude of* $f(x)$, is obtained by multiplying Eq. (12.18) by $e^{-ik'x}$ and integrating over x,

$$\int_{-\infty}^{\infty} f(x) e^{-ik'x} \, dx = \frac{1}{\sqrt{2\pi}} \int_{-\infty}^{\infty} g(k) \int_{-\infty}^{\infty} e^{i(k-k')x} \, dx dk$$

$$= \sqrt{2\pi} \int_{-\infty}^{\infty} g(k) \delta(k - k') \, dk,$$

---

[5]See Exercise 12.2.

where we have reversed the order of integration and made use of Eq. (12.17). Thus, the Fourier amplitude of $f(x)$ is seen to be

$$g(k) = \frac{1}{\sqrt{2\pi}} \int_{-\infty}^{\infty} f(x)e^{-ikx}\,dx. \tag{12.19}$$

As an example we obtain the Fourier expansion of the function

$$f(x) = \frac{1}{x^2 + a^2}, \tag{12.20}$$

where a is a real, positive number. The Fourier amplitude is

$$g(k) = \frac{1}{\sqrt{2\pi}} \int_{-\infty}^{\infty} \frac{e^{-ikx}}{x^2 + a^2}\,dx.$$

To evaluate this integral by contour integration we note that the integrand has simple poles at $z = \pm ia$. For $k < 0$ we complete the contour in the upper half-plane as shown in Figure 12.3a and apply Jordan's lemma to obtain

$$g(k) = \sqrt{\frac{\pi}{2}}\frac{e^{ka}}{a}, \qquad k < 0.$$

For $k > 0$ the contour is completed in the lower half-plane as in Figure 12.3b. In this case

$$g(k) = \sqrt{\frac{\pi}{2}}\frac{e^{-ka}}{a}, \qquad k > 0.$$

Now substitute these values for $g(k)$ into Eq. (12.18). After some rearrangement we get

$$f(x) = \frac{1}{a} \int_{0}^{\infty} e^{-ka}\cos kx\,dk. \tag{12.21}$$

from which Eq. (12.20) follows by direct integration.

Clearly, one virtue of dealing with complete sets of *orthonormal* functions lies in the ease with which the expansion coefficients are obtained when arbitrary functions are expressed as linear superpositions of the functions that are members of one of these complete sets.

12.3.1   AN APPLICATION IN QUANTUM MECHANICS

A quantity that can be measured is called an *observable*. In quantum mechanics an observable is represented by an operator whose eigenvalues are real and whose eigenstates form a complete set.[6] The linear momentum is

---

[6]P.A.M. Dirac, *op. cit.*, p. 37.

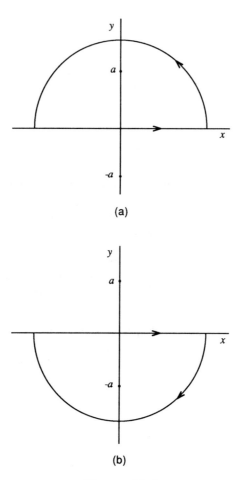

Figure 12.3

an observable. In the Schrödinger representation the one-dimensional momentum eigenvalue equation is

$$-i\hbar \frac{d}{dx} u(x,p) = pu(x,p),$$

where p is the momentum eigenvalue. On integrating we find that the (unnormalized) momentum eigenfunctions are

$$u(x,p) = e^{ipx/\hbar}.$$

Since these functions form a complete set, we have for an arbitrary function $\psi(x)$

$$\psi(x) = \int_{-\infty}^{\infty} c(p) e^{ipx/\hbar} dp. \tag{12.22}$$

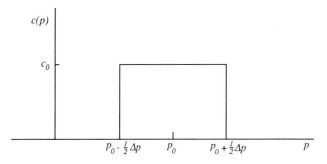

Figure 12.4

Note that $\sqrt{2\pi\hbar}c(p)$ is the Fourier amplitude of $\psi(x)$. That is,

$$c(p) = \frac{1}{2\pi\hbar} \int_{-\infty}^{\infty} \psi(x)e^{-ipx/\hbar}\,dx. \tag{12.23}$$

For example, consider a wave packet for which $c(p)$ is a constant over a momentum range Δp centered at p_0 and zero outside this range as illustrated in Figure 12.4. Specifically,

$$c(p) = \begin{cases} c_0, & p_0 - \frac{1}{2}\Delta p \leq p \leq p_0 + \frac{1}{2}\Delta p, \\ \\ 0, & \text{otherwise.} \end{cases}$$

On substituting this amplitude into Eq. (12.22) we get

$$\psi(x) = 2\hbar c_0 \frac{e^{ip_0 x/\hbar}}{x} \sin\left(\frac{x\Delta p}{2\hbar}\right). \tag{12.24}$$

The real and imaginary parts of this function are plotted in Figure 12.5.

12.3.2 MULTIDIMENSIONAL DELTA FUNCTIONS

The results of Section 12.3.1 are readily extended to any number of dimensions. For example, the Fourier integral of $f(\mathbf{r})$, a function of the three position coordinates, is

$$f(\mathbf{r}) = \frac{1}{(2\pi)^{3/2}} \int g(\mathbf{k})e^{i\mathbf{k}\cdot\mathbf{r}}\,d^3k.$$

By following the procedure above we find that the Fourier transform for $f(\mathbf{r})$ is

$$g(\mathbf{k}) = \frac{1}{(2\pi)^{3/2}} \int f(\mathbf{r})e^{-i\mathbf{k}\cdot\mathbf{r}}\,d^3r,$$

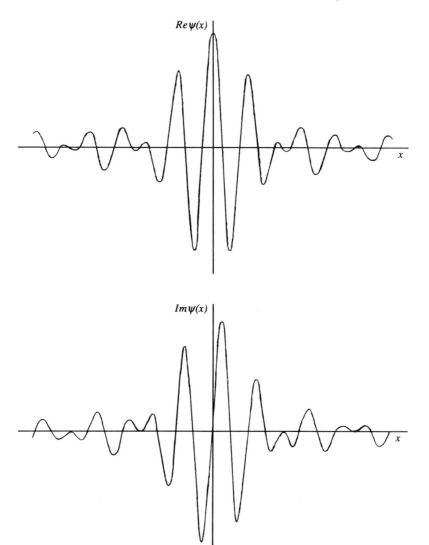

Figure 12.5

where we have used

$$\int e^{i(\mathbf{r}-\mathbf{r}')\cdot\mathbf{k}} d^3k = (2\pi)^3 \delta^3(\mathbf{r}-\mathbf{r}').$$

This latter expression is a generalization of Eq. (12.17) to three dimensions.

In rectangular coordinates

$$\delta^3(\mathbf{r} - \mathbf{r}') = \delta(x - x')\delta(y - y')\delta(z - z').$$

12.4 Orthogonality and Normalization of Special Functions

12.4.1 HERMITE POLYNOMIALS

With the results of Section 12.3 we can obtain an integral representation for the Hermite polynomials from the generating function

$$e^{x^2}e^{-(t-x)^2} = \sum_{n=0}^{\infty} \frac{H_n(x)}{n!} t^n. \tag{12.25}$$

We have the Fourier expansion[7]

$$e^{-(t-x)^2} = \frac{1}{2\sqrt{\pi}} \int_{-\infty}^{\infty} e^{-\frac{1}{4}k^2} e^{ik(t-x)} \, dk.$$

On writing the factor e^{ikt} in the integrand here as a Taylor series, the left-hand side of Eq. (12.25) can be expressed as

$$e^{x^2}e^{-(t-x)^2} = \frac{1}{2\sqrt{\pi}} \sum_{n=0}^{\infty} \frac{i^n t^n}{n!} \int_{-\infty}^{\infty} k^n e^{-\frac{1}{4}(k+2ix)^2} \, dk.$$

Now substitute this result into Eq. (12.25). After some rearrangement we get

$$\sum_{n=0}^{\infty} \frac{t^n}{n!} \left[H_n(x) - \frac{i^n}{2\sqrt{\pi}} \int_{-\infty}^{\infty} k^n e^{-\frac{1}{4}(k+2ix)^2} \, dk \right] = 0.$$

This equation holds for arbitrary values of t; therefore, coefficients of different powers of t must vanish separately. This gives the integral representation,

$$H_n(x) = \frac{i^n}{2\sqrt{\pi}} \int_{-\infty}^{\infty} k^n e^{-\frac{1}{4}(k+2ix)^2} \, dk. \tag{12.26}$$

The orthogonality of the harmonic oscillator eigenfunctions of Section 3.2.2 may be established from properties of the Hermite polynomials. If we substitute from Eq. (3.16) into Eq. (12.1), it becomes clear that the integral of interest is

$$\int_{-\infty}^{\infty} H_n(z) H_m(z) e^{-z^2} \, dz.$$

---

[7]See Exercise 12.6.

From the generating function in Eq. (12.25) we see that

$$e^{-z^2}e^{z^2-(u-z)^2}e^{z^2-(v-z)^2} = e^{-z^2}\sum_{n=0}^{\infty}\frac{H_n(z)}{n!}u^n\sum_{m=0}^{\infty}\frac{H_m(z)}{m!}v^m.$$

Let us integrate both sides of this equation over the full range of z. We get

$$\int_{-\infty}^{\infty}e^{-z^2+2(u+v)z-u^2-v^2}\,dz$$

$$= \sum_{n=0}^{\infty}\sum_{m=0}^{\infty}\frac{u^n v^m}{n!m!}\int_{-\infty}^{\infty}H_n(z)H_m(z)e^{-z^2}\,dz. \qquad (12.27)$$

Now, in the integral on the left complete the square in the exponent to obtain

$$\int_{-\infty}^{\infty}e^{-(z-u-v)^2+2uv}\,dz = \sqrt{\pi}e^{2uv}.$$

The Taylor series expansion of this exponential may be written as

$$\sqrt{\pi}e^{2uv} = \sqrt{\pi}\sum_{n=0}^{\infty}\sum_{m=0}^{\infty}\frac{2^n}{m!}u^n v^m\,\delta_{mn}.$$

Substitute this last expression for the integral on the left-hand side of Eq. (12.27) and collect terms. The equation then reads

$$\sum_{n=0}^{\infty}\sum_{m=0}^{\infty}\left[\frac{1}{n!}\int_{-\infty}^{\infty}H_n(z)H_m(z)e^{-z^2}\,dz - \sqrt{\pi}\,2^n\,\delta_{mn}\right]\frac{u^n v^m}{m!} = 0.$$

This equation holds for arbitrary values of u and v, which vary independently; therefore, the coefficients of $u^n v^m$ with different combinations of m and n must vanish separately, and the Hermite polynomials have the property

$$\int_{-\infty}^{\infty}H_m(z)H_n(z)e^{-z^2}\,dz = \sqrt{\pi}\,2^n n!\,\delta_{mn}. \qquad (12.28)$$

Clearly, the functions

$$u_n(z) = \pi^{-\frac{1}{4}}2^{-\frac{1}{2}n}(n!)^{-\frac{1}{2}}H_n(z)e^{-\frac{1}{2}z^2}$$

obey Eq. (12.1).

Finally, from Eq. (3.16) we see that the normalized one-dimensional harmonic oscillator eigenfunctions are

$$u_n(x) = \pi^{-\frac{1}{4}}2^{-\frac{1}{2}n}(bn!)^{-\frac{1}{2}}H_n(x/b)e^{-x^2/2b^2} \qquad (12.29)$$

12.4.2 BESSEL FUNCTIONS

Now let us establish the orthogonality of the functions

$$w_{mn}(\rho, \theta) = J_m\left(x_{mn}\frac{\rho}{b}\right)e^{im\theta}, \tag{12.30}$$

which arose in our analysis of the vibrating membrane.[8] By direct integration

$$\int_0^{2\pi} e^{i(m-m')\theta}\,d\theta = 2\pi\delta_{mm'}.$$

To show the orthogonality of the Bessel functions write

$$\left[\frac{1}{\rho}\frac{d}{d\rho}\left(\rho\frac{d}{d\rho}\right) + \left(\frac{x_{\nu n}^2}{b^2} - \frac{\nu^2}{\rho^2}\right)\right]J_\nu\left(x_{\nu n}\frac{\rho}{b}\right) = 0$$

and

$$\left[\frac{1}{\rho}\frac{d}{d\rho}\left(\rho\frac{d}{d\rho}\right) + \left(\frac{x_{\nu n'}^2}{b^2} - \frac{\nu^2}{\rho^2}\right)\right]J_\nu\left(x_{\nu n'}\frac{\rho}{b}\right) = 0,$$

both of which are Bessel's equations. We multiply the first of these by $\rho J_\nu\left(x_{\nu n'}\frac{\rho}{b}\right)$ and the second by $\rho J_\nu\left(x_{\nu n}\frac{\rho}{b}\right)$. Then, we integrate over ρ from 0 to b and subtract the second equation from the first to obtain

$$\int_0^b\left\{J_\nu\left(x_{\nu n'}\frac{\rho}{b}\right)\left[\frac{d}{d\rho}\left(\rho\frac{d}{d\rho}J_\nu\left(x_{\nu n}\frac{\rho}{b}\right)\right)\right]\right.$$
$$\left. - J_\nu\left(x_{\nu n}\frac{\rho}{b}\right)\left[\frac{d}{d\rho}\left(\rho\frac{d}{d\rho}J_\nu\left(x_{\nu n'}\frac{\rho}{b}\right)\right)\right]\right\}d\rho$$
$$+ \frac{x_{\nu n}^2 - x_{\nu n'}^2}{b^2}\int_0^b J_\nu\left(x_{\nu n'}\frac{\rho}{b}\right)J_\nu\left(x_{\nu n}\frac{\rho}{b}\right)\rho\,d\rho = 0.$$

On integrating the first integral by parts we find that

$$\left[J_\nu\left(x_{\nu n'}\frac{\rho}{b}\right)\rho\frac{d}{d\rho}J_\nu\left(x_{\nu n}\frac{\rho}{b}\right) - J_\nu\left(x_{\nu n}\frac{\rho}{b}\right)\rho\frac{d}{d\rho}J_\nu\left(x_{\nu n'}\frac{\rho}{b}\right)\right]\Bigg|_0^b$$
$$+ \frac{x_{\nu n}^2 - x_{\nu n'}^2}{b^2}\int_0^b J_\nu\left(x_{\nu n'}\frac{\rho}{b}\right)J_\nu\left(x_{\nu n}\frac{\rho}{b}\right)\rho\,d\rho = 0.$$

The term in square brackets vanishes at *both* limits for any $\nu \geq -1$. Thus,

$$\int_0^b J_\nu\left(x_{\nu n'}\frac{\rho}{b}\right)J_\nu\left(x_{\nu n}\frac{\rho}{b}\right)\rho\,d\rho = 0 \quad \text{for} \quad n' \neq n.$$

On combining this result with Eq. (4.39) we have

$$\int_0^b J_\nu\left(x_{\nu n'}\frac{\rho}{b}\right)J_\nu\left(x_{\nu n}\frac{\rho}{b}\right)\rho\,d\rho = \delta_{nn'}\frac{b^2}{2}J_{\nu\pm1}^2(x_{\nu n}), \tag{12.31}$$

---

[8]See Eq. (4.38).

for any $\nu \geq -1$.

Now write the orthogonality relation for the functions $w_{mn}(\rho, \theta)$ defined in Eq. (12.30) as,

$$\int_0^b \int_0^{2\pi} w^*_{m'n'}(\rho, \theta) w_{mn}(\rho, \theta) \, d\theta \rho \, d\rho$$

$$= \delta_{mm'} \delta_{nn'} \pi b^2 J^2_{m\pm 1}(x_{mn}). \tag{12.32}$$

This allows the construction of the complete solutions to the problem of the vibrating membrane discussed in Section 4.2. Because the functions $w_{mn}(\rho, \theta)$ form a complete set in the two-dimensional space of the polar coordinates ρ and θ, the solution to Eq. (4.30) can be written as

$$u(\rho, \theta, t) = \sum_{m=-\infty}^{\infty} \sum_{n=1}^{\infty} J_m \left(x_{mn} \frac{\rho}{b} \right) e^{im\theta} \left[a_{mn} \cos \omega_{mn} t + b_{mn} \sin \omega_{mn} t \right]. \tag{12.33}$$

The eigenfrequencies ω_{mn} are determined by the zeros of the Bessel functions according to

$$\omega_{mn} = \frac{v}{b} x_{mn}.$$

Here v is the propagation velocity[9] of the wave and b is the radius of the membrane. The coefficients a_{mn} and b_{mn} in Eq. (12.33) are obtained from the initial conditions for the membrane. These are given by Eqs. (4.35) and (4.36). For example, if $t = 0$ in Eq. (12.33) as required by Eq. (4.35), then

$$f(\rho, \theta) = \sum_{m=-\infty}^{\infty} \sum_{n=1}^{\infty} a_{mn} J_m \left(x_{mn} \frac{\rho}{b} \right) e^{im\theta}. \tag{12.34}$$

By multiplying both sides of this equation by $\rho J_{m'} \left(x_{m'n'} \frac{\rho}{b} \right) e^{-im'\theta}$ and integrating over ρ and θ

$$a_{mn} = \frac{1}{\pi b^2 J^2_{m\pm 1}(x_{mn})} \int_0^b \int_0^{2\pi} f(\rho, \theta) J_m \left(x_{mn} \frac{\rho}{b} \right) e^{-im\theta} \, d\theta \rho \, d\rho \tag{12.35}$$

is obtained. In a similar fashion according to Eq. (4.36) we find that

$$b_{mn} = \frac{1}{\pi v b x_{mn} J^2_{m\pm 1}(x_{mn})} \int_0^b \int_0^{2\pi} g(\rho, \theta) J_m \left(x_{mn} \frac{\rho}{b} \right) e^{-im\theta} \, d\theta \rho \, d\rho. \tag{12.36}$$

---

[9]This velocity depends on the density of the membrane and on how tightly it is stretched. See A.L. Fetter and J.D. Walecka, *op. cit.*, p. 273.

12.4.3 ASSOCIATED LAGUERRE FUNCTIONS

The radial wave functions $R_{nl}(r)$ for the isotropic harmonic oscillator are given in Eq. (6.62). The complete eigenfunctions $R_{nl}(r)Y_l^m(\theta, \phi)$ are considered to be orthonormal; therefore,

$$\int_0^\infty R_{nl}(r)R_{n'l}(r)r^2 dr = \delta_{nn'}. \qquad (12.37)$$

This leads to the appropriate orthogonality relation for the Laguerre polynomials. By substituting from Eq. (6.62) into the integral in Eq. (12.37) and changing the variable to $x = m\omega r^2/\hbar$ we find that

$$\int_0^\infty R_{nl}(r)R_{n'l}(r)r^2 dr$$

$$= \frac{N_{nl}N_{n'l}}{2}\left(\frac{\hbar}{m\omega}\right)^{3/2}\int_0^\infty e^{-x}x^{l+\frac{1}{2}}L_{n+l+\frac{1}{2}}^{l+\frac{1}{2}}(x)L_{n'+l+\frac{1}{2}}^{l+\frac{1}{2}}(x)\,dx.$$

Clearly, we are interested in the integral

$$\int_0^\infty x^p e^{-x}L_q^p(x)L_{q'}^p(x)\,dx. \qquad (12.38)$$

In the case of the oscillator q and p are not integers.[10] Nevertheless, the orthogonality relation for the associated Laguerre functions $L_q^p(x)$ takes the same form whether q and p are integers or not. Therefore, we use the generating function to find the orthogonality condition for integer q and p, then generalize to noninteger values.

The form of the integrand in Eq. (12.38) suggests using the generating function in Eq. (11.18) to write

$$x^p e^{-x}\left[(-1)^p\frac{u^p}{(1-u)^{p+1}}e^{-xu/(1-u)}\right]\left[(-1)^p\frac{v^p}{(1-v)^{p+1}}e^{-xv/(1-v)}\right]$$

$$= \sum_{j=0}^\infty\sum_{k=0}^\infty\frac{u^j v^k}{j!k!}x^p e^{-x}L_j^p(x)L_k^p(x).$$

We integrate both sides over the full range of x to obtain

$$\frac{\Gamma(p+1)(uv)^p}{(1-uv)^{p+1}} = \sum_{j=0}^\infty\sum_{k=0}^\infty\frac{u^j v^k}{j!k!}\int_0^\infty x^p e^{-x}L_j^p(x)L_k^p(x)\,dx. \qquad (12.39)$$

In carrying out this integration Eq. (1.16) was used to show that

$$\int_0^\infty x^p e^{-x(1+u/(1-u)+v/(1-v))}\,dx = \Gamma(p+1)\left(1+\frac{u}{1-u}+\frac{v}{1-v}\right)^{-p-1}.$$

---

[10]Note that their difference $q - p$ *is* a nonnegative integer. Therefore, from Eq. (6.47), it is clear that $L_q^p(x)$ is a polynomial.

If the denominator on the left-hand side of Eq. (12.39) is expanded as in Eq. (2.4), then Eq. (12.39) becomes

$$\Gamma(p+1)\sum_{s=0}^{\infty}\frac{(p+1)_s}{s!}(uv)^{s+p}$$

$$=\sum_{j=0}^{\infty}\sum_{k=0}^{\infty}\frac{u^j v^k}{j!k!}\int_0^{\infty}x^p e^{-x}L_j^p(x)L_k^p(x)\,dx. \tag{12.40}$$

Now, $L_j^p(x)=0$ for $j<p$, therefore, we can change the lower limits on the sums on the right-hand side of Eq. (12.40) to p without affecting the sums. Next, redefine the summation index on the left-hand side to be $j=s+p$. This gives

$$\Gamma(p+1)\sum_{j=p}^{\infty}\frac{(p+1)_{j-p}}{(j-p)!}(uv)^j=\sum_{j=p}^{\infty}\sum_{k=p}^{\infty}\frac{u^j v^k}{j!k!}\int_0^{\infty}x^p e^{-x}L_j^p(x)L_k^p(x)\,dx.$$

Now use the definition of the Kronecker delta to rewrite the left-hand side of this equation. After some rearrangement of terms the equation then reads

$$\sum_{j=p}^{\infty}\sum_{k=p}^{\infty}u^j v^k\left[\frac{1}{j!k!}\int_0^{\infty}x^p e^{-x}L_j^p(x)L_k^p(x)\,dx\right.$$

$$\left.-\frac{\Gamma(p+1)(p+1)_{k-p}}{(k-p)!}\delta_{jk}\right]=0.$$

Because u and v are arbitrary and independent, the coefficients of $u^j v^k$ for different combinations of j and k must vanish separately. Thus,

$$\int_0^{\infty}x^p e^{-x}L_j^p(x)L_k^p(x)\,dx=\frac{(k!)^2 p!(p+1)_{k-p}}{(k-p)!}\delta_{jk}.$$

With $k-p$ equal to an integer $\Gamma(p+1)(p+1)_{k-p}=\Gamma(k+1)$. Therefore, for the more general case where $q-p$ is an integer, but q and p are not necessarily integers,[11]

$$\int_0^{\infty}x^p e^{-x}L_q^p(x)L_{q'}^p(x)\,dx=\delta_{qq'}\frac{[\Gamma(q+1)]^3}{(q-p)!}. \tag{12.41}$$

In a similar fashion evaluate the integral

$$\int_0^{\infty}x^{p+1}e^{-x}[L_m^p(x)]^2\,dx,$$

---

[11]Note that if $q'\neq q$, then Eq. (12.41) cannot be used for the radial Coulomb eigenfunctions, because the argument of $L_q^p(x)$ is *not* the same as that of $L_{q'}^p(x)$. See Eq. (6.49).

which will be useful in normalizing the bound state Coulomb eigenfunctions[12] of Section 6.4.2. From the generating function another relation similar to Eq. (12.39) is obtained,

$$(1 - u)(1 - v) \sum_{k=0}^{\infty} \frac{(p + k + 1)!}{k!} (uv)^{k+p}$$

$$= \sum_{n=0}^{\infty} \sum_{m=0}^{\infty} \frac{u^n v^m}{n! m!} \int_0^{\infty} x^{p+1} L_n^p(x) L_m^p(x) e^{-x} \, dx.$$

Since u and v are arbitrary and independent, coefficients of $u^n v^m$ for any given m and n must be the same on both sides of this equation. In particular if $m = n$,

$$\frac{(m + 1)!}{(m - p)!} + \frac{m!}{(m - p - 1)!} = \frac{1}{(m!)^2} \int_0^{\infty} x^{p+1} e^{-x} [L_m^p(x)]^2 \, dx.$$

From this result the value for the integral is seen to be

$$\int_0^{\infty} x^{p+1} e^{-x} [L_q^p(x)]^2 \, dx = \frac{[\Gamma(q + 1)]^3}{(q - p)!} (2q + 1 - p). \tag{12.42}$$

12.4.4 ASSOCIATED LEGENDRE FUNCTIONS

The associated Legendre functions are orthogonal according to

$$\int_{-1}^{1} P_l^m(x) P_{l'}^m(x) \, dx = 0 \quad \text{for} \quad l \neq l'. \tag{12.43}$$

We use Rodrigues's formula (Eq. 9.21) to prove this result.[13]

$$\int_{-1}^{1} P_l^m(x) P_{l'}^m(x) \, dx$$

$$= \frac{(-1)^m}{2^{l+l'} l! l'!} \int_{-1}^{1} (x^2 - 1)^m \frac{d^{m+l}}{dx^{m+l}} (x^2 - 1)^l \frac{d^{m+l'}}{dx^{m+l'}} (x^2 - 1)^{l'} \, dx.$$

For definiteness take $l' > l$ and integrate by parts $m + l'$ times to get

---

[12] See Exercise 12.13.

[13] As a notational convenience here we drop the vertical bars on $|m|$. We restore them in the final result.

$$\int_{-1}^{1} P_l^m(x) P_{l'}^m(x) \, dx$$

$$= \frac{(-1)^m}{2^{l+l'} l! l'} \sum_{s=0}^{m+l'-1} (-1)^s \left[\frac{d^s}{dx^s} \left((x^2-1)^m \frac{d^{m+l}}{dx^{m+l}} (x^2-1)^l \right) \frac{d^{m+l'-s-1}}{dx^{m+l'-s-1}} (x^2-1)^{l'} \right] \Bigg|_{-1}^{1}$$

$$+ \frac{(-1)^{l'}}{2^{l+l'} l! l'!} \int_{-1}^{1} \frac{d^{m+l'}}{dx^{m+l'}} \left[(x^2-1)^m \frac{d^{m+l}}{dx^{m+l}} (x^2-1)^l \right] (x^2-1)^{l'} \, dx. \quad (12.44)$$

By writing $(x^2-1)^k = (x-1)^k (x+1)^k$ we can see from Leibniz's theorem that

$$\frac{d^p}{dx^p} (x^2-1)^k = (-1)^p p! \sum_{q=0}^{p} \frac{(-k)_q (-k)_{p-q}}{q!(p-q)!} (x-1)^{k-q} (x+1)^{k+q-p}.$$

If $x = +1$, then only the term with $q = k$ contributes to the sum. Similarly, if $x = -1$, then only the term with $q = p - k$ contributes. For $p < k$, both of these terms lie outside the range of q. Thus,

$$\frac{d^p}{dx^p} (x^2-1)^k \Bigg|_{-1}^{1} = 0 \quad \text{for} \quad p < k. \quad (12.45)$$

On using this result in Eq. (12.44) those terms in the sum with $m + l' - s - 1 < l'$ vanish. Therefore, for nonvanishing terms in the sum we must have $s < m$.

Application of Leibniz's theorem to the other factor in the sum in Eq. (12.44) leads to

$$\frac{d^s}{dx^s} \left[(x^2-1)^m \frac{d^{m+l}}{dx^{m+l}} (x^2-1)^l \right]$$

$$= \sum_{r=0}^{s} \frac{s!}{r!(s-r)!} \frac{d^r}{dx^r} (x^2-1)^m \frac{d^{m+l+s-r}}{dx^{m+l+s-r}} (x^2-1)^l.$$

But according to Eq. (12.45),

$$\frac{d^r}{dx^r} (x^2-1)^m \Bigg|_{-1}^{1} = 0$$

for all r, because $r \le s < m$. In Eq. (12.44) only the integral on the right-hand side remains. The function in the square brackets in the integrand is a polynomial in x of degree $m + l$. This function is differentiated $m + l'$ times to get

$$\frac{d^{m+l'}}{dx^{m+l'}} \left[(x^2-1)^m \frac{d^{m+l}}{dx^{m+l}} (x^2-1)^l \right] = 0,$$

since $l' > l$. Thus, Eq. (12.43) is established.

If $l' = l$, again only the term with the integral in Eq. (12.44) survives. In the integrand,

$$\frac{d^{m+l}}{dx^{m+l}} \left[(x^2 - 1)^m \frac{d^{m+l}}{dx^{m+l}} (x^2 - 1)^l \right]$$

$$= \frac{d^{m+l}}{dx^{m+l}} \left[(x^2 - 1)^m \left((l - m + 1)_{l+m} x^{l-m} + \cdots \right) \right]$$

$$= \frac{(2l)!(l+m)!}{(l-m)!}$$

Now substitute this result into Eq. (12.44) with $l' = l$ to get[14]

$$\int_{-1}^{1} P_l^m(x) p_l^m(x)\, dx = \frac{(-1)^l}{2^{2l} l! l!} \frac{(2l)!(l + |m|)!}{(l - |m|)!} \int_{-1}^{1} (x^2 - 1)^l\, dx.$$

By a change of the variable of integration according to $x = 2t - 1$ we obtain for the integral

$$\int_{-1}^{1} (x^2 - 1)^l\, dx = (-1)^l 2^{2l+1} \int_{0}^{1} (1 - t)^l t^l\, dt$$

$$= (-1)^l 2^{2l+1} \frac{l! l!}{(2l+1)!},$$

where we have used Eq. (1.22). Thus,

$$\int_{-1}^{1} P_l^m(x) P_l^m(x)\, dx = \frac{2}{2l+1} \frac{(l + |m|)!}{(l - |m|)!}. \tag{12.46}$$

Finally, on combining Eqs. (12.43) and (12.46) we have

$$\int_{-1}^{1} P_l^m(x) P_{l'}^m(x)\, dx = \frac{2}{2l+1} \frac{(l + |m|)!}{(l - |m|)!} \delta_{ll'}. \tag{12.47}$$

12.4.5 SPHERICAL HARMONICS

We are now in a position to define the orthonormal set of complex functions $Y_l^m(\theta, \phi)$ in Eq. (5.26), which are solutions of the angular part of the Schrödinger equation for a spherically symmetric force field. Require that

$$\int Y_l^{m*}(\theta, \phi) Y_{l'}^{m'}(\theta, \phi)\, d\Omega = \delta_{ll'} \delta_{mm'}. \tag{12.48}$$

The solid angle $d\Omega = \sin\theta\, d\theta d\phi$. Figure 5.1 shows that the range on θ is $0 \leq \theta \leq \pi$ and on ϕ is $0 \leq \phi \leq 2\pi$. First, the ϕ integration in Eq. (12.48) gives

$$\int_{0}^{2\pi} e^{i(m'-m)\phi}\, d\phi = 2\pi \delta_{mm'}.$$

---

[14]The vertical bars in $|m|$ have now been restored.

Then, for the θ integration,

$$C^*_{lm} C_{l'm} \int_0^\pi P_l^m(\cos\theta) P_{l'}^m(\cos\theta) \sin\theta \, d\theta = C^*_{lm} C_{l'm} \int_{-1}^1 P_l^m(x) P_{l'}^m(x) \, dx,$$

$$= |C_{lm}|^2 \frac{2}{2l+1} \frac{(l+|m|)!}{(l-|m|)!} \delta_{ll'},$$

where $x = \cos\theta$ and Eq. (12.47) is invoked.

The requirement in Eq. (12.48) is met if

$$|C_{lm}| = \left[\frac{2l+1}{4\pi} \frac{(l-|m|)!}{(l+|m|)!}\right]^{1/2}.$$

It is conventional[15] to choose the phase of C_{lm} such that

$$Y_l^m(\theta,\phi) = i^{m+|m|} \left[\frac{2l+1}{4\pi} \frac{(l-|m|)!}{(l+|m|)!}\right]^{1/2} e^{im\phi} P_l^m(\cos\theta). \qquad (12.49)$$

These are the *spherical harmonics*. Clearly, they have the property

$$Y_l^{m*}(\theta,\phi) = (-1)^m Y_l^{-m}(\theta,\phi). \qquad (12.50)$$

12.5 Applications of Spherical Harmonics

12.5.1 RAYLEIGH EXPANSION OF A PLANE WAVE

From Eq. (10.13) the spherical Bessel function can be written as

$$j_l(y) = (-1)^l \frac{y^l}{l! 2^{l+1}} \int_{-1}^1 e^{iyx}(x^2-1)^l \, dx. \qquad (12.51)$$

Differentiation l times of e^{iyx} with respect to x gives

$$\frac{d^l}{dx^l} e^{iyx} = i^l y^l e^{iyx}. \qquad (12.52)$$

With this result Eq. (12.51) becomes

$$j_l(y) = (-1)^l \frac{i^{-l}}{l! 2^{l+1}} \int_{-1}^1 \left[\frac{d^l}{dx^l} e^{iyx}\right](x^2-1)^l \, dx.$$

Now integrate by parts l times to get

$$j_l(y) = \frac{i^{-l}}{l! 2^{l+1}} \int_{-1}^1 e^{iyx} \frac{d^l}{dx^l}(x^2-1)^l \, dx,$$

---

[15]E.U. Condon and G.H. Shortley, *Theory of Atomic Spectra*, Cambridge University Press, Cambridge, 1935.

where the integral contains the Rodrigues representation of the Legendre polynomial (see Eq. 9.20). Thus,

$$j_l(y) = \frac{i^{-l}}{2} \int_{-1}^{1} e^{iyx} P_l(x)\, dx. \qquad (12.53)$$

A plane wave of wavelength $2\pi/k$ propagating along the z axis is described in spherical polar coordinates (see Figure 5.1) by

$$e^{ikz} = e^{ikr\cos\theta}.$$

Solutions to the free-particle Schrödinger equation are given in Eq. (6.28). Those with symmetry about the polar axis $(m = 0)$ are

$$j_l(kr)P_l(\cos\theta).$$

Because these eigenfunctions must form a complete set, the plane wave can be written as a linear superposition of them, that is,

$$e^{ikr\cos\theta} = \sum_{l=0}^{\infty} c_l j_l(kr)P_l(\cos\theta).$$

Set $x = \cos\theta$, multiply by $P_{l'}(x)$, and integrate over x to get

$$c_l = (2l+1)i^l.$$

In obtaining this result Eq. (12.53) and the orthogonality relation for the Legendre functions (Eq. 12.47) have been used. Thus,

$$e^{ikr\cos\theta} = \sum_{l=0}^{\infty} (2l+1)i^l j_l(kr)P_l(\cos\theta). \qquad (12.54)$$

This is the *Rayleigh expansion*[16] of the plane wave.

12.5.2 ELECTRIC POTENTIAL OUTSIDE AN ARBITRARY DISTRIBUTION OF CHARGE

Now look at an example from electrostatics. If there are no time-varying fields in the region outside the source, two of Maxwell's equations (Eqs. 4.1a and 4.1b) reduce to

$$\nabla \cdot \mathbf{E} = 0 \qquad (12.55)$$

and

$$\nabla \times \mathbf{E} = 0. \qquad (12.56)$$

---

[16]Also called the *partial wave* expansion.

If we let $\mathbf{E}$ be equal to the gradient of some scalar function U,

$$\mathbf{E} = -\nabla U, \tag{12.57}$$

then Eq. (12.56) is satisfied automatically, since the curl of the gradient of *any* scalar function vanishes identically. Substitute from Eq. (12.57) into Eq. (12.55) to see that U is the solution to the equation

$$\nabla^2 U = 0. \tag{12.58}$$

This is *Laplace's equation*. Because of the separability of the radial and angular parts of the operator ∇^2 in spherical polar coordinates (see Eq. 5.3), the solutions are products

$$U(r, \theta, \phi) = R(r)Y(\theta, \phi).$$

This product with the operator from Eq. (5.3) inserted into Eq. (12.58) leads to the equation

$$\frac{1}{R(r)}\frac{\partial}{\partial r}\left(r^2\frac{\partial}{\partial r}\right)R(r)$$

$$= \frac{-1}{Y(\theta,\phi)}\left(\frac{1}{\sin\theta}\frac{\partial}{\partial\theta}\left(\sin\theta\frac{\partial}{\partial\theta}\right) + \frac{1}{\sin^2\theta}\frac{\partial^2}{\partial\phi^2}\right)Y(\theta,\phi).$$

Both sides must be equal to the same constant λ. As we saw in solving Eq. (5.8), to obtain physically acceptable solutions the separation constant must be of the form

$$\lambda = n(n+1),$$

where n is a nonnegative integer. The angular solutions are the spherical harmonics. The radial part of the solution satisfies the differential equation

$$r^2\frac{d^2}{dr^2}R(r) + 2r\frac{d}{dr}R(r) - n(n+1)R(r) = 0.$$

Solutions of this equation are of the form r^p. By direct substitution we find that $p = \pm\left(n + \frac{1}{2}\right) - \frac{1}{2}$. Thus, the complete solution to Eq. (12.58) is

$$U(r, \theta, \phi) = \sum_{n=0}^{\infty}\sum_{m=-n}^{n}\left(A_{nm}r^n + B_{nm}r^{-n-1}\right)Y_n^m(\theta, \phi). \tag{12.59}$$

As an application consider an uncharged conducting sphere of radius R placed in a uniform electric field $\mathbf{E}_0$. We choose the origin of the coordinate system to be at the center of the sphere with the positive z axis in the direction of the uniform field. The electric potential is described by Laplace's equation (Eq. 12.58). With symmetry about the z axis the potential does not depend on ϕ and Eq. (12.59) reduces to

$$U(r, \theta) = \sum_{n=0}^{\infty}\left(a_n r^n + b_n r^{-n-1}\right)P_n(\cos\theta). \tag{12.60}$$

To determine the coefficients a_n and b_n, invoke the following boundary conditions:

a. The electric field at the surface of the conductor must be perpendicular to the surface. Otherwise, there will be electric currents in the conductor.

$$E_\theta\big|_{r=R} = 0 \qquad \text{or} \qquad \frac{\partial U}{\partial \theta}\bigg|_{r=R} = 0.$$

b. At large distances from the sphere the field lines heal and the electric field is uniform,

$$\mathbf{E} \xrightarrow[r \to \infty]{} \mathbf{e}_z E_0,$$

where $\mathbf{e}_z$ is a unit vector.

According to Eq. (12.57) this last boundary condition is equivalent to the requirement

$$-\frac{\partial U}{\partial z} = E_0,$$

which we integrate to obtain

$$U \xrightarrow[r \to \infty]{} -E_0 r \cos \theta + U_0,$$

where U_0 is a constant. On comparing this result with Eq. (12.60) in this same limit, we find that

$$a_0 = U_0,$$
$$a_1 = -E_0,$$
$$a_n = 0 \qquad \text{for} \ \ n \geq 2.$$

Thus,

$$U(r, \theta) = \left(U_0 + \frac{b_0}{r} \right) P_0(\cos \theta) + \left(\frac{b_1}{r^2} - E_0 r \right) P_1(\cos \theta)$$

$$+ \sum_{n=2}^{\infty} \frac{b_n}{r^{n+1}} P_n(\cos \theta). \tag{12.61}$$

The boundary condition at the surface of the sphere requires that U be independent of θ there. Because the Legendre polynomials are linearly independent, it follows from Eq. (12.61) that

$$b_0 = 0, \qquad \text{sphere is uncharged,}$$
$$b_1 = E_0 R^3,$$
$$b_n = 0 \qquad \text{for} \ \ n \geq 2.$$

The complete solution for $r \geq R$ is

$$U(r, \theta) = U_0 + E_0 r \left(\frac{R^3}{r^3} - 1 \right) P_1(\cos \theta), \qquad r \geq R.$$

With this result the components of the electric field outside the sphere can be calculated. These are

$$E_r = -\frac{\partial U}{\partial r} = E_0 \left(\frac{2R^3}{r^3} + 1 \right) \cos \theta$$

and

$$E_\theta = -\frac{1}{r} \frac{\partial U}{\partial \theta} = E_0 \left(\frac{R^3}{r^3} - 1 \right) \sin \theta,$$

which satisfy the boundary conditions.

12.6   Sturm–Liouville Theory

An alternate development of the theory of orthogonal functions is provided by the *Sturm–Liouville theory*. It is useful to see how that theory is related to the track followed in this book. For this purpose it will be sufficient to restrict our discussion to real orthogonal functions.

First, note that the special functions of a single variable that we have considered obey orthogonality relations with the general form

$$\int_a^b u_{\lambda'}(x) u_\lambda(x) w(x) \, dx = 0 \quad \text{for} \quad \lambda' \neq \lambda. \tag{12.62}$$

Here $w(x)$ is a weighting function which is real. Note that the labeling of the eigenfunction $u_\lambda(x)$ with its eigenvalue λ (also real) is purely formal. In practice the label is usually some number related to the eigenvalue. The Legendre polynomials $P_l(x)$, for example, correspond to eigenvalues given by $\lambda = l(l+1)$.

The eigenvalue equation is a linear, second-order differential equation,

$$a(x)u''(x) + b(x)u'(x) + c(x)u(x) = 0. \tag{12.63}$$

Presently, we shall show that this equation can also be written as

$$\Omega u_\lambda(x) + \lambda w(x) u_\lambda(x) = 0, \tag{12.64}$$

where Ω is a differential operator. By multiplying Eq. (12.64) by $u_{\lambda'}(x)$ and integrating from a to b we see that

$$\int_a^b u_{\lambda'}(x) \Omega u_\lambda(x) \, dx = -\lambda \int_a^b u_{\lambda'}(x) u_\lambda(x) w(x) \, dx. \tag{12.65}$$

Now rewrite Eq. (12.64) for λ', multiply by $u_\lambda(x)$, and integrate over the same range. This gives

$$\int_a^b (\Omega u_{\lambda'}(x)) u_\lambda(x)\, dx = -\lambda' \int_a^b u_{\lambda'}(x) u_\lambda(x) w(x)\, dx. \qquad (12.66)$$

Subtract Eq. (12.66) from Eq. (12.65) to get

$$\int_a^b u_{\lambda'}(x) \Omega u_\lambda(x)\, dx - \int_a^b (\Omega u_{\lambda'}(x)) u_\lambda(x)\, dx$$
$$= (\lambda' - \lambda) \int_a^b u_{\lambda'}(x) u_\lambda(x) w(x)\, dx.$$

If for $\lambda' \neq \lambda$ the functions $u_{\lambda'}(x)$ and $u_\lambda(x)$ are orthogonal according to Eq. (12.62), then clearly

$$\int_a^b u_{\lambda'}(x) \Omega u_\lambda(x)\, dx = \int_a^b (\Omega u_{\lambda'}(x)) u_\lambda(x)\, dx. \qquad (12.67)$$

Any operator Ω that has this property is said to be *self-adjoint* (or *hermitian*).

Equation (12.63) can be expressed in the form of Eq. (12.64) by writing

$$\frac{d}{dx}\left[f(x) \frac{d}{dx} u(x) + g(x) u(x) \right] + \lambda h(x) u(x) = 0$$

and choosing

$$f(x) = a(x),$$
$$g(x) = -a'(x) + b(x),$$
$$h(x) = \lambda^{-1}\left[a''(x) - b'(x) + c(x) \right].$$

A more restricted form of Eq. (12.63) is given by

$$\frac{d}{dx}\left[s(x) \frac{d}{dx} u(x) \right] + t(x) u(x) + \lambda w(x) u(x) = 0 \qquad (12.68)$$

with

$$s(x) = a(x),$$
$$s'(x) = b(x),$$
$$t(x) + \lambda w(x) = c(x).$$

A differential equation of this latter type is called a *Sturm–Liouville equation*. For Sturm–Liouville equations the operator Ω has the form

$$\Omega = \frac{d}{dx}\left[s(x) \frac{d}{dx} \right] + t(x). \qquad (12.69)$$

We now show that it is operators of this kind that have the property required by Eq. (12.67).

Using the operator in Eq. (12.69), we find that two integrations by parts give

$$\int_a^b u_{\lambda'} \Omega u_\lambda dx = u_{\lambda'} \left(\frac{d}{dx} u_\lambda \right) s \Big|_a^b - \left(\frac{d}{dx} u_{\lambda'} \right) u_\lambda s \Big|_a^b + \int_a^b (\Omega u_{\lambda'}) u_\lambda \, dx$$

from which we see that Ω is self-adjoint if

$$u_{\lambda'}(x) u_\lambda'(x) s(x) \Big|_a^b = u_{\lambda'}'(x) u_\lambda(x) s(x) \Big|_a^b .$$

For the special functions that we have considered this condition is met, because the product $u_\alpha'(x) u_\beta(x) s(x)$ has the same value at both limits for all α and β.

If a differential equation can be cast into the Sturm–Liouville form, then the orthogonality of solutions is assured provided certain rather mild boundary conditions are satisfied. From the differential equations for the special functions it is a simple matter to construct the Sturm–Liouville equation and identify its components. For example, Laguerre's equation is

$$xL_q''(x) + (1 - x)L_q'(x) + qL_q(x) = 0. \tag{12.70}$$

Clearly, $s(x)$ in Eq. (12.68) must have the form

$$s(x) = xg(x)$$

with the constraint that

$$s'(x) = xg'(x) + g(x) = (1 - x)g(x).$$

This last equation can be integrated directly for $g(x)$ to obtain

$$s(x) = xe^{-x}.$$

Now substitute this expression into Eq. (12.68) and compare the result with Eq. (12.70); then

$$\lambda = q, \qquad t(x) = 0, \qquad w(x) = e^{-x}.$$

Bessel's equation requires a slightly different treatment, because in the orthogonality relation the eigenvalue label appears in the argument of the function. In Eq. (12.31) set $x = \rho/b$ and write the orthogonality relation for Bessel functions as

$$\int_0^1 J_\nu(\mu' x) J_\nu(\mu x) x \, dx = 0 \quad \text{for} \quad \mu' \neq \mu.$$

The parameters μ and μ' denote zeros of $J_\nu(y)$, which satisfies

$$y^2 J_\nu''(y) + y J_\nu'(y) + (y^2 - \nu^2) J_\nu(y) = 0.$$

With the change of variable $y = \mu x$, we have

$$x^2 \frac{d^2}{dx^2} J_\nu(\mu x) + x \frac{d}{dx} J_\nu(\mu x) + (\mu^2 x^2 - \nu^2) J_\nu(\mu x) = 0.$$

Now, following the procedure above for the Laguerre equation, expressing Bessel's equation in Sturm–Liouville form requires

$$s(x) = x, \qquad t(x) = -\nu^2/x, \qquad \lambda = \mu^2, \qquad w(x) = x.$$

The elements of the Sturm-Liouville equations for the other special functions that we have considered in this book are given in Table 12.1.

TABLE 12.1. Elements of the Sturm–Liouville equation (Eq. 12.68) for special functions

Special Function	$s(x)$	$t(x)$	$w(x)$	λ
Hermite polynomial	e^{-x^2}	0	e^{-x^2}	$2n$
Bessel	x	$-\nu^2/x$	x	μ^2
Legendre polynomial	$1 - x^2$	0	1	$l(l+1)$
Associated Legendre	$1 - x^2$	$-m(1 - x^2)^{-1}$	1	$l(l+1)$
Chebyshev Type 1	$(1 - x^2)^{1/2}$	0	$(1 - x^2)^{-1/2}$	n^2
Chebyshev Type 2	$(1 - x^2)^{3/2}$	0	$(1 - x^2)^{1/2}$	$n(n+2)$
Laguerre polynomial	xe^{-x}	0	e^{-x}	q
Associated Laguerre	$x^{p+1}e^{-x}$	0	$x^p e^{-x}$	$q - p$
Confluent hypergeo- metric	$x^c e^{-x}$	0	$x^{c-1}e^{-x}$	$-a$

EXERCISES

1. Calculate the area bounded by the curve and the t axis for each of the functions $y(t, p)$ depicted in Figure 12.1.

2. Show that
$$\int_{-\infty}^{\infty} \frac{\sin pt}{t}\, dt = -i \int_{-\infty}^{\infty} \frac{e^{ipt}}{t}\, dt.$$

   Apply Jordan's lemma to the integral on the right-hand side to obtain the Cauchy principal value,
$$P \int_{-\infty}^{\infty} \frac{\sin pt}{t}\, dt = \pi.$$

   Hint: Use the contour shown in Figure 12.6.

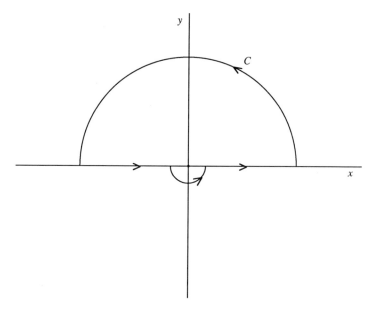

Figure 12.6

3. The function $f(x) = (x^2 + a^2)^{-1}$ is given in Eq. (12.20). By means of contour integration evaluate the Fourier amplitude $g(k)$ for this function when $k = 0$. Show that your result is consistent with the values of $g(k)$ obtained in the text for $k \neq 0$.

4. Carry out the integration in Eq. (12.21) to obtain the original form of $f(x)$ given in Eq. (12.20).

5. Calculate the Fourier amplitudes for the function $(x^4 + 4b^4)^{-1}$ and obtain the integral representation

$$\frac{1}{x^4 + 4b^4} = \frac{1}{4b^3} \int_0^\infty e^{-kb}(\cos kb + \sin kb) \cos kx \, dk.$$

6. Calculate the Fourier amplitudes for the Gaussian function $e^{-x^2/2b^2}$ and show that

$$e^{-x^2/2b^2} = \frac{b}{\sqrt{2\pi}} \int_{-\infty}^\infty e^{-\frac{1}{2}b^2 k^2} e^{ikx} \, dk.$$

7. Find c_0 in Eq. (12.24) such that the function $\psi(x)$ representing the wave packet is normalized to unity. Show that

$$\psi(x) = \sqrt{\frac{2\hbar}{\pi \Delta p}} \frac{e^{ip_0 x/\hbar}}{x} \sin\left(\frac{x \Delta p}{2\hbar}\right).$$

8. The Fourier integral for any potential energy function $V(\mathbf{r})$ may be written as

$$V(\mathbf{r}) = \frac{1}{(2\pi)^{3/2}} \int U(\mathbf{q}) e^{i\mathbf{q} \cdot \mathbf{r}} d^3 q.$$

Show that for any central force field the Fourier transform is

$$U(\mathbf{q}) = U(q) = \sqrt{\frac{2}{\pi}} \int_0^\infty V(r) j_0(qr) r^2 \, dr.$$

9. Find the central field potential energy function whose Fourier transform is

$$U(q) = \sqrt{\frac{2}{\pi}} \frac{A}{q^2 + m^2},$$

where A and m are constants. Show that this transform follows from the result obtained in the previous exercise.[17]

10. The function

$$y(x) = \frac{(\Gamma/2)^2}{(x - x_0)^2 + \frac{1}{4}\Gamma^2}$$

is plotted in Figure 12.7. It is seen that Γ is the width of the curve at one-half the maximum height. Calculate the Fourier amplitude of this

---

[17]This is the potential energy function in Yukawa's meson theory of the nuclear force. It is called the *Yukawa potential.*

function. Plot the real and imaginary parts of the Fourier amplitude as a function of $k\Gamma$. Show that

$$y(x) = \frac{\Gamma}{2} \int_0^\infty e^{-\frac{1}{2}\Gamma k} \cos[k(x - x_0)]\, dk.$$

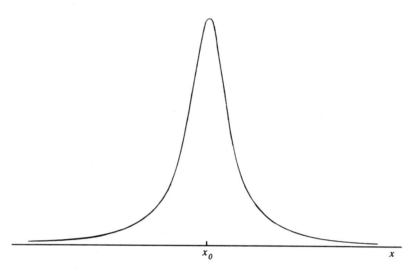

Figure 12.7

11. The radial wave function for the quantum harmonic oscillator is normalized such that

$$\int_0^\infty [R_{nl}(r)]^2 r^2\, dr = 1.$$

Calculate the normalization constant N_{nl} for this function given in Eq. (6.62).

12. Fill in the missing steps in the derivation of the normalization integral of Eq. (12.42),

$$\int_0^\infty x^{p+1} e^{-x} [L_q^p(x)]^2\, dx = \frac{[\Gamma(q + 1)]^3}{(q - p)!}(2q + 1 - p).$$

13. Calculate the normalization constant N_{nl} for the Coulomb radial wave function of Eq. (6.48).

14. For an electron bound in the Coulomb field of a proton, the average value of any physical observable A for the electron in a given eigenstate is

$$\overline{A} = \int u_{nlm}^*(\mathbf{r}) \hat{A} u_{nlm}(\mathbf{r}) \, d^3r,$$

where $\hat{A}$ is the operator corresponding to A. Calculate the electrostatic potential energy of the electron in any eigenstate. Compare this with the *total* energy of the electron in this state. How much kinetic energy does it have in this state?

15. Use the orthogonality relation for Hermite polynomials to evaluate

$$\int_{-\infty}^{\infty} H_n(x) e^{-x^2} \, dx.$$

16. The integral in Eq. (12.26) may be written as

$$\int_{-\infty}^{\infty} k^n e^{-\frac{1}{4}(k+2ix)^2} \, dk = e^{x^2} T_n\left(\tfrac{1}{4}, ix\right)$$

where the function $T_n(a, b)$ is defined by

$$T_n(a, b) = \int_{-\infty}^{\infty} t^n e^{-at^2 - bt} \, dt.$$

By making a change of variable $s = t + b/2a$, show that

$$T_n(a, b) = (-1)^n \sqrt{\frac{\pi}{a}} \frac{e^{b^2/4a}}{2^n} \left(\frac{b}{a}\right)^n \sum_{m=0}^{[n/2]} \frac{(-n)_{2m}}{m!} \left(\frac{a}{b^2}\right)^m.$$

The binomial expansion will be useful here.

17. With the result in the previous exercise show that the expression for $H_n(x)$ given in Eq. (12.26) reduces to the series representation of $H_n(x)$ in Eq. (3.15).

18. The mean-square displacement from equilibrium of a particle in the nth state of a one-dimensional quantum oscillator is

$$\overline{x^2} = \int_{-\infty}^{\infty} u_n(x) x^2 u_n(x) \, dx.$$

Find $\overline{x^2}$ in terms of the length b of Eq. (12.29). You will find the recursion formula for the Hermite polynomials useful in this problem.

19. For a one-dimensional system the kinetic energy operator is

$$\frac{-\hbar^2}{2m}\frac{d^2}{dx^2}.$$

    Calculate the average kinetic energy of the particle in the previous exercise.

20. Carry out explicitly the calculation of the coefficients a_{mn} and b_{mn} from the initial conditions on the solution to the vibrating membrane problem given in Eq. (12.33).

21. Use the eigenfunctions for the isotropic harmonic oscillator to calculate the mean square displacement from equilibrium of a particle in the state represented by $u_{nlm}(\mathbf{r})$.

22. The wave function of a certain particle constrained to move along the x-axis is

$$\psi(x) = \frac{2}{\pi^{1/4}\sqrt{3bb^2}}\, x^2 e^{-x^2/2b^2}.$$

    This function can be written as a superposition of momentum eigenfunctions according to

$$\psi(x) = \int_{-\infty}^{\infty} \phi(p)e^{ipx/\hbar}\, dp.$$

    Calculate $\phi(p)$ which gives the momentum distribution in $\psi(x)$.

23. A particle of mass m is confined to a two-dimensional circular region of radius R. From Eq. (4.4) we see that in this case the time-independent Schrödinger equation takes the form

$$\frac{-\hbar^2}{2m}\left[\frac{1}{\rho}\frac{\partial}{\partial\rho}\left(\rho\frac{\partial}{\partial\rho}\right) + \frac{1}{\rho^2}\frac{\partial^2}{\partial\theta^2}\right]u(\rho,\theta) + V(\rho,\theta)u(\rho,\theta) = Eu(\rho,\theta),$$

    where the potential energy is given by[18]

$$V(\rho,\theta) = \begin{cases} 0, & \rho \le R, \\ \infty, & \rho > R. \end{cases}$$

    Invoke the boundary conditions on the solutions to Schrödinger's equation at $\rho = 0$ and $\rho = R$ to obtain the energy spectrum and the normalized energy eigenfunctions for this particle.

---

[18]V.I. Kogan and V.M. Galitskiy,*Problems in Quantum Mechanics*, Prentice-Hall, Englewood Cliffs, NJ, 1963, p. 3.

Bibliography

G. Arfken, *Mathematical Methods for Physicists*, Academic Press, London, 1970.

G.F. Carrier, M. Krook, and C.E. Pearson, *Functions of a Complex Variable*, McGraw-Hill, New York, 1966.

R.V. Churchill, *Introduction to Complex Variables and Applications*, McGraw-Hill, New York, 1948.

E.U. Condon and G.H. Shortley, *Theory of Atomic Spectra*, Cambridge University Press, Cambridge, 1935.

P.A.M. Dirac, *The Principles of Quantum Mechanics*, Oxford University Press, Oxford, 1958.

A. Erdélyi, Ed., *Higher Transcendental Functions*, vol 1–3, McGraw-Hill, New York, 1955.

A.L. Fetter and J.D. Walecka, *Theoretical Mechanics of Particles and Continua*, McGraw-Hill, New York, 1980.

P. Franklin, *Functions of Complex Variables*, Prentice-Hall, Englewood Cliffs, NJ, 1958.

D.J. Griffiths, *Introduction to Electrodynamics*, Prentice-Hall, Englewood Cliffs, NJ, 1981.

D. ter Haar, *Selected Problems in Quantum Mechanics*, Academic Press, New York, 1964.

E. Hecht and A. Zajac, *Optics*, Addison-Wesley, Reading, MA, 1974.

J.D. Jackson, *Classical Electrodynamics*, Wiley, New York, 1962.

H. Jeffreys and B.S. Jeffreys, *Methods of Mathematical Physics*, Cambridge University Press, Cambridge, 1956.

W. Kaplan, *Advanced Calculus*, Addison-Wesley, Reading, MA, 1953.

C. Kittel, *Introduction to Solid State Physics*, Wiley, New York, 1956.

V.I. Kogan and V.M. Galitskiy, *Problems in Quantum Mechanics*, Prentice-Hall, Englewood Cliffs, NJ, 1963.

E.A. Kraut, *Fundamentals of Mathematical Physics*, McGraw-Hill, New York, 1972.

N.N. Lebedev, *Special Functions and Their Applications*, Dover, New York, 1972.

P. Lorraine and D. Corson, *Electromagnetic Fields and Waves*, W.H. Freeman, San Francisco, 1970.

H. Margenau and G. Murphy, *The Mathematics of Physics and Chemistry*, Van Nostrand, Princeton, 1956.

J. Mathews and R. Walker, *Mathematical Methods of Physics*, Benjamin, New York, 1965.

P.T. Matthews, *Introduction to Quantum Mechanics*, McGraw-Hill, London, 1968.

E. Merzbacher, *Quantum Mechanics*, Wiley, New York, 1961.

P.M. Morse and H. Feshbach, *Methods of Theoretical Physics*, McGraw-Hill, 1953.

J.F. Randolph, *Calculus*, Macmillan, New York, 1952.

F. Reif, *Fundamentals of Statistical and Thermal Physics*, McGraw-Hill, New York, 1965.

L.I. Schiff, *Quantum Mechanics*, McGraw-Hill, New York, 1955.

B. Spain and G.M. Smith, *Functions of Mathematical Physics*, Van Nostrand Reinhold, London, 1970.

M.R. Spiegel, *Complex Variables*, Schaum's Outline Series, McGraw-Hill, New York, 1964.

E.C. Titchmarsh, *The Theory of Functions*, Oxford University Press, Oxford, 1939.

G.N. Watson, *A Treatise on the Theory of Bessel Functions*, Cambridge University Press, Cambridge, 1922.

E.T. Whittaker and G.N. Watson, *A Course in Modern Analysis*, Cambridge University Press, Cambridge, 1927.

R.G. Winter, *Quantum Mechanics*, Wadsworth, Belmont, CA 1979.

Index